Electrostatic Discharge Protection

Advances and Applications

T0227840

Devices, Circuits, and Systems

Series Editor

Krzysztof Iniewski
CMOS Emerging Technologies Research Inc.,
Vancouver, British Columbia, Canada

PUBLISHED TITLES:

Atomic Nanoscale Technology in the Nuclear Industry
Taeho Woo

Biological and Medical Sensor Technologies
Krzysztof Iniewski

Building Sensor Networks: From Design to Applications
Ioanis Nikolaidis and Krzysztof Iniewski

**Cell and Material Interface: Advances in Tissue Engineering,
Biosensor, Implant, and Imaging Technologies**
Nihal Engin Vrana

Circuits at the Nanoscale: Communications, Imaging, and Sensing
Krzysztof Iniewski

CMOS: Front-End Electronics for Radiation Sensors
Angelo Rivetti

Design of 3D Integrated Circuits and Systems
Rohit Sharma

Electrical Solitons: Theory, Design, and Applications
David Ricketts and Donhee Ham

Electronics for Radiation Detection
Krzysztof Iniewski

Electrostatic Discharge Protection: Advances and Applications
Juin J. Liou

**Embedded and Networking Systems:
Design, Software, and Implementation**
Gul N. Khan and Krzysztof Iniewski

Energy Harvesting with Functional Materials and Microsystems
Madhu Bhaskaran, Sharath Sriram, and Krzysztof Iniewski

Gallium Nitride (GaN): Physics, Devices, and Technology
Farid Medjdoub

**Graphene, Carbon Nanotubes, and Nanostuctures:
Techniques and Applications**
James E. Morris and Krzysztof Iniewski

PUBLISHED TITLES:

FORTHCOMING TITLES:

Power Management Integrated Circuits and Technologies
Mona M. Hella and Patrick Mercier

Radio Frequency Integrated Circuit Design
Sebastian Magierowski

Silicon on Insulator System Design
Bastien Giraud

Semiconductor Devices in Harsh Conditions
Kirsten Weide-Zaage and Malgorzata Chrzanowska-Jeske

Smart eHealth and eCare Technologies Handbook
Sari Merilampi, Lars T. Berger, and Andrew Sirkka

Structural Health Monitoring of Composite Structures Using Fiber Optic Methods
Ginu Rajan and Gangadhara Prusty

Terahertz Sensing and Imaging: Technology and Devices
Daryoosh Saeedkia and Wojciech Knap

Tunable RF Components and Circuits: Applications in Mobile Handsets
Jeffrey L. Hilbert

Wireless Medical Systems and Algorithms: Design and Applications
Pietro Salvo and Miguel Hernandez-Silveira

Electrostatic Discharge Protection

Advances and Applications

EDITED BY
JUIN J. LIOU
University of Central Florida, Orlando, USA

KRZYSZTOF INIEWSKI MANAGING EDITOR
CMOS Emerging Technologies Research Inc.
Vancouver, British Columbia, Canada

CRC Press
Taylor & Francis Group
Boca Raton London New York

CRC Press is an imprint of the
Taylor & Francis Group, an **informa** business

CRC Press
Taylor & Francis Group
6000 Broken Sound Parkway NW, Suite 300
Boca Raton, FL 33487-2742

First issued in paperback 2017

© 2016 by Taylor & Francis Group, LLC
CRC Press is an imprint of Taylor & Francis Group, an Informa business

No claim to original U.S. Government works

ISBN-13: 978-1-4822-5588-1 (hbk)
ISBN-13: 978-1-138-89307-8 (pbk)

This book contains information obtained from authentic and highly regarded sources. Reasonable efforts have been made to publish reliable data and information, but the author and publisher cannot assume responsibility for the validity of all materials or the consequences of their use. The authors and publishers have attempted to trace the copyright holders of all material reproduced in this publication and apologize to copyright holders if permission to publish in this form has not been obtained. If any copyright material has not been acknowledged please write and let us know so we may rectify in any future reprint.

Except as permitted under U.S. Copyright Law, no part of this book may be reprinted, reproduced, transmitted, or utilized in any form by any electronic, mechanical, or other means, now known or hereafter invented, including photocopying, microfilming, and recording, or in any information storage or retrieval system, without written permission from the publishers.

For permission to photocopy or use material electronically from this work, please access www.copyright.com (http://www.copyright.com/) or contact the Copyright Clearance Center, Inc. (CCC), 222 Rosewood Drive, Danvers, MA 01923, 978-750-8400. CCC is a not-for-profit organization that provides licenses and registration for a variety of users. For organizations that have been granted a photocopy license by the CCC, a separate system of payment has been arranged.

Trademark Notice: Product or corporate names may be trademarks or registered trademarks, and are used only for identification and explanation without intent to infringe.

Visit the Taylor & Francis Web site at
http://www.taylorandfrancis.com

and the CRC Press Web site at
http://www.crcpress.com

Contents

Chapter 10 Compact Modeling of Semiconductor Devices for Electrostatic

Zhenghao Gan and Waisum Wong

Chapter 11

Vladislav A. Vashchenko and Andrei A. Shibkov

Chapter 12

David L. Catlett, Jr., Roger A. Cline, and Ponnarith Pok

Chapter 13 ESD Design and Optimization in Advanced CMOS SOI

You Li

Preface

Electrostatic discharge (ESD) is one of the most prevalent threats to the reliability of electronic components. It is an event in which a finite amount of charge is transferred from one object (i.e., human body) to another (i.e., microchip). This process can result in a very high current passing through the microchip within a very short period of time, and, hence, more than 35% of chip damages can be attributed to an ESD-related event. As such, designing on-chip ESD structures to protect integrated circuits against the ESD stresses is a high priority in the semiconductor industry. The continuing advancement in metal oxide semiconductor and other processing technologies makes ESD-induced failures even more prominent. In fact, many semiconductor companies worldwide are having difficulties in meeting the increasingly stringent ESD protection requirements for various electronics applications, and one can predict with certainty that the availability of effective and robust ESD protection solutions will become a critical and essential factor to the well-being and commercialization of modern and future electronic devices.

This book contains timely, comprehensive, informative, and up-to-date materials on the topic of ESD protection for semiconductor devices and/or integrated circuits. It is excellently structured to bridge the gap between theory and practice and illustrated amply with tables, figures, and case studies. The book consists of 13 chapters, and for the first time it brings together a team of experienced and well-respected researchers and engineers around the world with expertise in ESD design, optimization, modeling, simulation, and characterization. Subjects relevant to component- and system-level ESD protection are covered; the technologies considered include the Si CMOS, Si BCD, Si SOI, and compound semiconductor processes. While the majority of chapters in the book offer valuable insights into ESD protection design and optimization, chapters 1 and 2 provide extensive introductory and background materials on ESD and chapters 10 and 11 focus on important aspects pertinent to the modeling and simulation of ESD protection solutions.

I thank all the contributing authors, who were devoted in preparing these excellent chapters despite their busy schedule and time constraints. I also express my appreciation to my wife, Peili; my son, Will; and my daughter, Monica, who have sacrificed their quality time with me due to my frequent absences from home to pursue my academic career, promote international collaboration, and engage in professional service.

Juin J. Liou
University of Central Florida

Editors

Juin J. Liou received his B.S. (honors), M.S., and Ph.D. degrees in electrical engineering from the University of Florida, Gainesville, in 1982, 1983, and 1987, respectively. In 1987, he joined the Department of Electrical and Computer Engineering at the University of Central Florida (UCF), Orlando, Florida, where he is now the Pegasus distinguished professor, Lockheed Martin St. Laurent professor, and UCF-Analog Devices fellow. His current research interests are micro/nanoelectronics computer-aided design; RF device modeling and simulation; and electrostatic discharge protection design, modeling, and simulation.

Dr. Liou holds 8 U.S. patents (5 more filed and pending), and he has published 10 books (3 more in press), more than 270 journal papers (including 18 invited review articles), and more than 220 papers (including 99 keynote and invited papers) in international and national conference proceedings. He has been awarded more than $14.0 million of research contracts and grants from federal agencies (i.e., NSF, DARPA, Navy, Air Force, NASA, and NIST), state government, and industry (i.e., Semiconductor Research Corporation, Intel Corporation, Intersil Corporation, Lucent Technologies, Alcatel Space, Conexant Systems, Texas Instruments, Fairchild Semiconductor, National Semiconductor, Analog Devices, Maxim Integrated Systems, Allegro Microsystems, RF Micro Device, and Lockheed Martin), and he has held consulting positions with research laboratories and companies in the United States, China, Japan, Taiwan, and Singapore. In addition, Dr. Liou has served as a technical reviewer for various journals and publishers, general chair or technical program chair for a large number of international conferences, regional editor (in United States, Canada, and South America) of the *Microelectronics Reliability* journal, and guest editor of six special issues in the *IEEE Journal on Emerging and Selected Topics in Circuits and Systems, Microelectronics Reliability, Solid-State Electronics, and International Journal of Antennas and Propagation.*

Dr. Liou has received ten different awards for excellence in teaching and research from the University of Central Florida (UCF) and six different awards from the IEEE. Among them, he was awarded the UCF Pegasus Distinguished Professor (2009)—the highest honor bestowed to a faculty member at UCF; UCF Distinguished Researcher Award (four times: 1992, 1998, 2002, and 2009)—the most of any faculty in the history of UCF; UCF Research Incentive Award (three times: 2000, 2005, and 2010); UCF Trustee Chair Professor (2002); IEEE Joseph M. Biedenbach Outstanding Engineering Educator Award in 2004 for exemplary engineering teaching, research, and international collaboration; and IEEE Electron Devices Society Education Award in 2014 for promoting and inspiring global education and learning in the field of electron devices. He is a fellow of IEEE, IET, Singapore Institute of Manufacturing Technology, and UCF-Analog Devices and a distinguished lecturer in IEEE Electron Device Society (EDS) and National Science Council. He holds several honorary professorships, including the Chang Jiang Scholar Endowed Professor of Ministry of Education, China—the highest honorary professorship in China; NSVL Distinguished Professor of National Semiconductor Corporation,

United States; International Honorary Chair Professor of National Taipei University of Technology, Taiwan; Chang Gung Endowed Professor of Chang Gung University, Taiwan; Feng Chia Chair Professor of Feng Chia University, Taiwan; Chunhui Eminent Scholar of Peking University, China; Cao Guang-Biao Endowed Professor of Zhejiang University, China; honorary professor of Xidian University, China; consultant professor of Huazhong University of Science and Technology, China; and courtesy professor of Shanghai Jiao Tong University, China. Dr. Liou was a recipient of U.S. Air Force Fellowship Award and National University Singapore Fellowship Award.

Dr. Liou has served as the IEEE EDS vice-president of regions/chapters, IEEE EDS treasurer, IEEE EDS Finance Committee chair, member of IEEE EDS Board of Governors, and member of IEEE EDS Educational Activities Committee.

Krzysztof ("Kris") Iniewski is managing research and development at Redlen Technologies, Inc., a start-up company in Vancouver, Canada. Redlen's revolutionary production process for advanced semiconductor materials enables a new generation of more accurate, all-digital, radiation-based imaging solutions. Kris is also president of CMOS Emerging Technologies Research, Inc. (www.cmosetr.com), which organizes high-tech events covering communications, microsystems, optoelectronics, and sensors. In his career, Kris has held numerous faculty and management positions at the University of Toronto, University of Alberta, SFU, and PMC-Sierra, Inc. He has published and presented over 100 research papers in international journals and conferences. He holds 18 international patents granted in the United States, Canada, France, Germany, and Japan. Kris is a frequent invited speaker and has been consulted by multiple organizations internationally. He has written and edited several books for CRC Press, Cambridge University Press, IEEE Press, Wiley, McGraw-Hill, Artech House, and Springer. His goal is to contribute to healthy living and sustainability through innovative engineering solutions. In his leisure time, Kris can be found hiking, sailing, skiing, or biking in beautiful British Columbia, Canada. He can be reached at kris.iniewski@gmail.com.

Contributors

David L. Catlett, Jr.
Texas Instruments
Dallas, Texas

Roger A. Cline
Texas Instruments
Dallas, Texas

Charvaka Duvvury
Texas Instruments (Retired)
Dallas, Texas

Zhenghao Gan
Semiconductor Manufacturing
    International Corporation
Shanghai, China

Jean-Jacques Hajjar
Analog Devices
Wilmington, Massachusetts

Ming-Dou Ker
National Chiao Tung University
Hsinchu, Taiwan

You Li
IBM
Essex Junction, Vermont

Juin J. Liou
University of Central Florida
Orlando, Florida

Kevin Mello
Qorvo
Greensboro, North Carolina

Gaudenzio Meneghesso
University of Padova
Padova, Italy

Matteo Meneghini
University of Padova
Padova, Italy

Nathaniel Peachey
Qorvo
Greensboro, North Carolina

Ponnarith Pok
Texas Instruments
Dallas, Texas

Alan W. Righter
Analog Devices
Wilmington, Massachusetts

Javier A. Salcedo
Analog Devices
Wilmington, Massachusetts

Andrei A. Shibkov
Angstrom DA
San Jose, California

Teruo Suzuki
Fujitsu Corporation
Tokyo, Japan

Vladislav A. Vashchenko
Maxim Integrated Systems
Palo Alto, California

Jim Vinson
Intersil Corporation
Melbourne, Florida

Waisum Wong
Semiconductor Manufacturing
    International Corporation
Shanghai, China

Chih-Ting Yeh
National Chiao Tung University
Hsinchu, Taiwan

Enrico Zanoni
University of Padova
Padova, Italy

1 Introduction to Electrostatic Discharge Protection

Juin J. Liou

CONTENTS

1.1 BACKGROUND OF ELECTROSTATIC DISCHARGE EVENTS

Electrostatic discharge (ESD) is one of the most prevalent threats to the integrity of electronic components and integrated circuits (ICs). It is an event in which a finite amount of charge is transferred from one object (i.e., human body) to another (i.e., microchip). This process can result in a very high current passing through the object within a very short period of time [1,2]. A common ESD phenomenon is shown in Figure 1.1, where a person is shocked by an ESD from the metal doorknob to the ground via the human body. In this case, the human body is the charge source, the metal knob is the ground, and the human body skin is the conducting path between the source and ground. The amount of charges on a body is typically described by the voltage, and the voltage level depends on factors such as flooring material and air humidity. When a microchip or an electronic system is subject to an ESD event, the huge ESD-induced current can damage the microchip and cause malfunction to the electronic system if the ESD-generated energy in the object cannot be dissipated quickly enough. Figure 1.2 shows various damages found in microchips resulted from ESD stresses. These damages can be summarized into three categories: oxide breakdown, junction failure, and metal fusing. It is estimated that about 35% of all damaged microchips are ESD related, resulting in a loss of revenue of several hundred million dollars in the global semiconductor industry every year [3]. The continuing scaling of complementary metal-oxide semiconductor (CMOS) technology makes ESD-induced failures even more prominent, and one can predict with certainty that the availability of effective and robust ESD protection solutions will become a critical and essential component to the advancement and commercialization of the modern and next-generation Si, GaAs, GaN, and other technologies [4–7].

In this chapter, the fundamentals of ESD, including its mechanisms, standards, protection design principles, and testing, will be briefly introduced.

1

FIGURE 1.1 A commonly known ESD phenomenon.

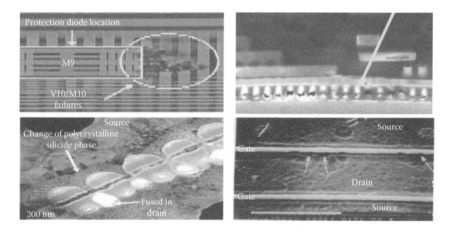

FIGURE 1.2 ESD-induced damages (clockwise, from the upper left-hand corner) associated with interconnect burnout, metal contact rupture, junction melting, and crystal structure change.

The following three processes can generate charges on an object: the triboelectrification process where the charges are generated because of the contacting and rubbing of two different objects, the inductive process where the charges are generated because of the presence of an electric field, and the conductive process where an uncharged object is charged after it is in touch with another charged object. The amount of charge generated increases with decreasing air humidity, and the severity of the charge is described by the voltage. For example, a human body can carry more than 10,000 V of ESD in a dry environment and a few hundred volts of ESD in a wet environment [8].

Various ESD events taking place in our daily life and in microchip manufacturing environment can be classified into four standards or models: the human body model (HBM), which describes an event when a charged person touches a microchip; the machine model (MM), which describes an event when a charged metal object is in contact with a microchip; the charged device model (CDM), which describes an event when a charged device is in contact with a grounded object; and the International Electrotechnical Commission (IEC), which describes an event when a charged cable/wire is in contact with an electronic component.

Some comments on the different ESD standards are in order. HBM is a mature, well-understood ESD model for simulating charge transfer from a person's finger to an electronic component. However, recent industry data indicates that HBM rarely simulates real-world ESD failures. Latest-generation package styles such as mBGAs, SOTs, SC70s, and CSPs with mm-range dimensions are often effectively too small for people to handle with fingers. Even in cases of relatively large components, most high-volume component and board manufacturing uses automated equipment, so humans rarely touch the components. CDM can more successfully replicate in-house and customer IC failures at the component level. It simulates the damage induced when a metal pin or solder ball on a charged IC package is instantaneously discharged via contact with a metallic object at ground potential.

An effective way to protect an electronics system against ESD events is to incorporate an ESD protection structure on the microchip (called the on-chip ESD protection) to increase the survivability of the core circuit when an ESD strikes. In such a structure, all input, output, and power supply pins of the core circuit are connected to the ground bus/rail via ESD protection devices. These ESD devices must be in the off-state during the normal system operation (i.e., in the absence of an ESD event), must be turned on quickly when an ESD event takes place so that the current generated by the ESD event can be conducted by ESD protection devices and discharged to the ground, must themselves not be damaged by the ESD stress, must clamp the pins to a sufficiently low voltage during the ESD event, and must return to the off-state after the ESD event has passed [9,10].

There are a number of technologies for which this on-chip ESD protection is not feasible and using an off-chip ESD protection solution becomes necessary. The off-chip ESD protection structure can be incorporated in electrical cables, in connectors, in ceramic carriers, or on circuit boards.

In the future, designing effective ESD protection solutions will become increasingly difficult and costly. As shown in Figure 1.3, the cost for designing ESD protection solutions in general increases with CMOS technology advancement (i.e., the technology node reducing from 180 nm to 45 nm) [11]. Furthermore, using custom solutions (ESD protection solutions designed with customized and optimized approaches) has the advantage of cutting the cost over using public solutions (ESD protection solutions designed with generalized and off-the-shelf approaches). The operation of an ESD protection device is described by the black curve in Figure 1.4. It is sandwiched between the IC operating area on the left and the IC reliability constraint on the right. As the CMOS technology is advancing, this design window becomes smaller, making the ESD protection design more challenging.

1.2 ESD PROTECTION DESIGN PRINCIPLES

As mentioned in the beginning of the chapter, effective ESD protection solutions are must-haves for modern ICs. The design of ESD protection solutions, however, is challenging and difficult because of the constraints imposed by the technology,

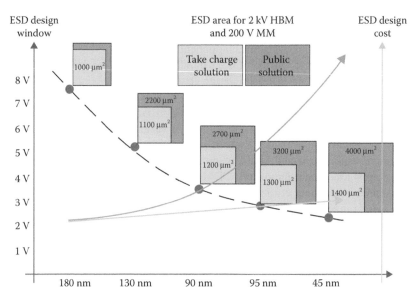

FIGURE 1.3 ESD design cost as a function of the CMOS technology using the custom solution (light gray line) and the public solution (dark gray line). (Data from Van Mele, K., *Effective ESD Strategies in Nano-CMOS IC Design*, Sarnoff Europe, Gistel, Belgium, http://www.chipestimate.com/techtalk.php?d=2007-12-04.)

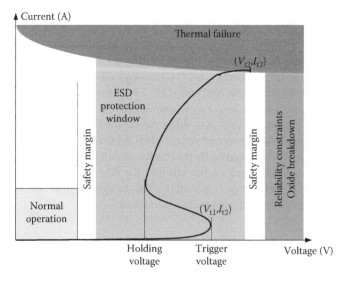

FIGURE 1.4 Illustration of IC operating area, ESD design window, I–V curve of ESD protection device, and IC reliability constraint.

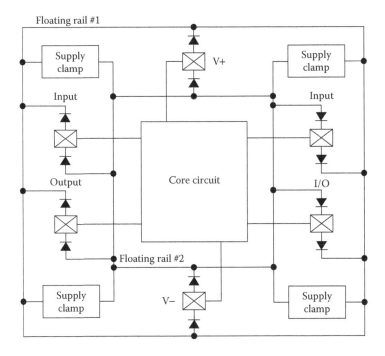

FIGURE 1.5 Typical ESD protection scheme constructed using diodes at the I/O pins and supply clamps between the power supply and ground bus.

IC operation, and customer demand. In this section, we will focus on issues pertinent to the design of on-chip ESD protection solutions.

A typical on-chip ESD protection scheme is shown in Figure 1.5, where two ESD protection devices (i.e., diodes) are connected to each input/output (I/O) pin and several ESD protection devices/circuits, called supply clamps, are connected between the power supply rail V_{dd} and ground bus V_{ss}. Such a scheme is used to protect the core circuit against positive- and negative-polarity ESD stresses by creating a conducting loop when any one pin is subject to an ESD stress and any other pin is grounded. The resulting conducting path, consisting of ESD I/O protection devices, supply clamps, and metal lines, renders the following two objectives during an ESD event: (1) it reduces the likelihood of ESD-induced current being entering the core circuit, and (2) it clamps the voltage of the protected pin to an acceptably low level. Both these objectives can therefore minimize the prospect of ESD-induced damages to the core circuit. It should be mentioned that, under an ESD event, the diodes are more suited to turn on and conduct the current under the forward-biased condition. As such, diodes are called one-directional devices, and each I/O pin would require the placement of two diodes (see Figure 1.5). On the other hand, only one element is needed for each I/O pin when employing bidirectional devices, such as silicon-controlled rectifiers (SCRs), because it can offer high robustness in conducting the current in both forward- and reverse-biased directions. More details on this subject will be given later. Another important consideration is the placement of supply clamps, which can be constructed with devices or circuits. Without supply clamps, some of the diodes will be forced to operate in the undesirable reverse-biased condition, and hence the sizes of the diodes must be enlarged to ensure that the ESD protection structure possesses a sufficient robustness. Multiple supply clamps around the V_{dd}/V_{ss} rails are also needed to minimize the distance, and thus the voltage drop on the current discharging route, between any two pins that are zapped and grounded.

Although the principle of ESD protection is fairly straightforward, several requirements must be considered and met before a protection solution can success-fully be designed and implemented [1,8–10]. They include the following:

1. The ESD protection device must be in the off-state during the normal system operation.
2. The ESD protection device must be turned on quickly when an ESD event occurs.
3. The resistance of the current discharging route associated with the ESD protection structure must be much lower than that of the current discharging path through the core circuit.
4. The voltage at the pin to which the ESD protection device is connected must be maintained at a sufficiently low value during the ESD event to avoid core circuit failures.
5. The ESD protection device as well as the core circuit cannot be damaged by the ESD stress.
6. The ESD protection device must return to the off-state after the ESD event has passed, otherwise devices will operate in the prohibited latch-up state.
7. Small size, low leakage current, high transparency, high robustness, and low cost are preferred.

ESD protection devices can be classified into non-snapback and snapback devices. Figure 1.6a and b shows the quasi-static current–voltage (I–V) characteristics of non-snapback and snapback ESD protection devices, respectively, operating under the ESD condition. Let us first discuss the snapback behavior depicted in Figure 1.6b. There are three important operating points in this device: the trigger point (point of device turn-on), holding point (point of device operation), and failure point (point of device being damaged). The voltage at the trigger point, called the trigger volt-age, V_{t1}, must be located within a voltage range called the ESD design window. The lower bound of the window is the operating voltage at the pin to which the ESD protection device is connected, and the upper bound is the maximum voltage the pin can tolerate without causing damages to the core circuit. The values of these two boundaries depend strongly on the type of core circuit being protected (digital vs. analog, low voltage vs. high voltage, etc.). The snapback mechanism reduces the voltage drop and thus the power dissipation on the protection device, hence result-ing in an increase in the robustness of the device. The voltage at the holding point, called the holding voltage, V_h, can influence the failure point. The smaller the hold-ing voltage, the larger the failure current, I_{t2} (current at the failure point). From this

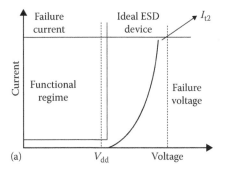

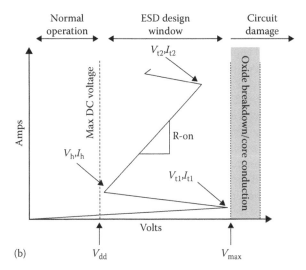

FIGURE 1.6 Quasi-static I–V curve of (a) non-snapback and (b) snapback ESD protection devices.

perspective, it is advantageous to reduce V_h of the protection device. However, having a too-small V_h can induce the risk of latch-up, unless V_h is larger than the operating voltage of the pin. As such, an ideal protection device would possess a holding voltage larger than the operating voltage of the pin but not too much larger to sacrifice the device's robustness. Moving upward from the holding point along the I–V curve in Figure 1.6b is the operating region of the ESD device. The curve's slope represents the on-state resistance, which should be minimized to reduce the likelihood of the voltage being larger than the upper-bound voltage discussed earlier and the ESD-induced damages to the core circuit. Finally, a high failure current I_{t2} is desirable, as it determines the robustness, or the ESD protection capability, of the ESD protection device. As mentioned earlier, I_{t2} is inversely proportional to V_h. In addition, a higher robustness can be achieved by increasing the device size but at the expenses of increasing Si consumption and increasing parasitic capacitance associated with the ESD protection device. All the concepts discussed earlier apply to a non-snapback device, except that in this case the trigger voltage is the same as the holding voltage, as shown in Figure 1.6a. The correlation between the holding point (i.e., holding voltage and holding current) and the latch-up needs to be clarified and elaborated. The reason an ESD protection device is latched up is because there is a sufficiently large current passing through this ESD device, and such a current keeps the device turned on even in the absence of the ESD stress. Two approaches can be used to eliminate the threat of latch-up. The first is to make sure that the ESD protection device's holding voltage is larger than the operating voltage at the pin. This potential difference will remove the possibility of pouring a large current from the pin to the ground via the ESD protection device. The second is to make the device's holding current surpassing the current available at the pin. For a supply clamp connected between V_{dd} and V_{ss}, the second approach would not work as the current available at V_{dd} is normally very large. Consequently, only the first approach can be implemented. For the ESD protection device connected to an I/O pin, on the other hand, the second approach is more feasible as the current level available at the I/O pin is typically very limited.

Passing voltage level is commonly used to gauge the ESD protection capability. For HBM, the maximum ESD protection a device can provide, in terms of the voltage, is the failure current multiplied to the human body resistance of 1500 Ω. For other ESD models, the correlation between the failure current and passing voltage level is not yet clearly established.

Three semiconductor devices are frequently used to realize ESD protection solutions: diode, grounded-gate n-channel MOSFET (GGNMOS), and SCR. Among them, the diode has the simplest structure and is more suited for low-voltage ESD applications because of its low trigger voltage in the forward-biased condition. The GGNMOS is widely used in CMOS-based ESD applications because of its familiar structure and operation. The SCR possesses the highest robustness per unit area because of its bipolar conduction mechanism. But these three devices have their own cons, including the high leakage current, low robustness, and proneness to latch-up for the diode, GGNMOS, and SCR, respectively.

The quasi-static I–V curves of SCR, GGNMOS, and diode operating under the ESD condition are compared in Figure 1.7. Clearly, the diode is a non-snapback-type device, whereas SCR and GGNMOS are snapback-type devices. All these devices can be triggered and conduct the current in both positive and negative directions. For the diode in the forward direction, the trigger and holding voltages are identical and about 0.7 V. This is quite good for the design of ESD protection for a low-voltage IC, as the trigger voltage required for such an application is relatively low. But for protecting a pin with a higher operating voltage, a few diodes would need to be connected in series to increase the trigger voltage (i.e., trigger voltage is number of diodes × 0.7 V) to be larger than the lower bound of the ESD design window. This has the disadvantages of consuming a large die area and increasing the parasitic resistance at the pin. A diode can also trigger and conduct current in the breakdown region under the negative direction. The trigger voltage under this operation is relatively large, so it is potentially useful for ESD protection of a high-voltage IC. But as the holding voltage is the same as the trigger voltage (no snapback), the diode used in this operation suffers a very low robustness. As a result, the diode is

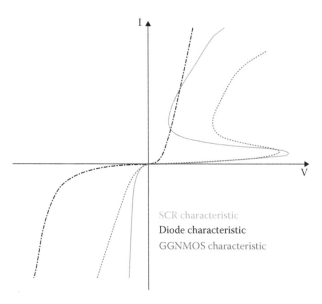

FIGURE 1.7 Quasi-static I–V curves of the diode, GGNMOS, and SCR under the ESD condition.

normally referred to as a one-directional device, which is suitable for use only in the positive ESD operation but not in the negative ESD operation. The GGNMOS and SCR, on the other hand, have a snapback behavior in the positive direction, and they exhibit a forward-biased diode characteristic that triggers at −0.7 V in the negative direction. Thus, these two devices can provide relatively versatile bidirectional ESD protection, but their large snapback window can sometimes impose a challenge to fit the ESD operation within the ESD design window. Another main drawback of the GGNMOS is its low robustness due to the fact that the current conducts near the surface in such a device. On the other hand, for the SCR, the current conduction takes place in the bulk and hence a high robustness.

It should be noted that the quasi-static I–V curves mentioned earlier are to be obtained from the transmission line pulsing (TLP) technique rather than from the conventional curve tracer. More details about the TLP testing are given in the next section.

1.3 ESD MEASUREMENT AND TESTING

The ESD protection structure testing typically involves two steps. The first step is to measure an ESD protection device using a TLP tester to examine the device's characteristics (i.e., trigger voltage, holding voltage, failure current, etc.) subject to an ESD stress. After the protection device is verified by the TLP tester and is considered as a good candidate for constructing the on-chip ESD protection solution, then the device is integrated with the core circuit. In the second testing step,

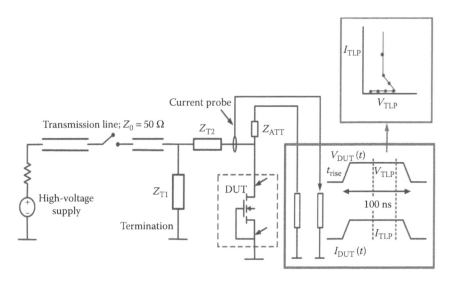

FIGURE 1.8 Diagram showing how the quasi-static I–V curve of a DUT can be generated under the TLP testing.

the microchip, now consisting of the core circuit and ESD protection structure, is zapped using an ESD tester to see if the microchip can pass a certain type and level of ESD stress.

The TLP tester is an equipment of paramount importance to ESD protection characterization. The main purpose of such an equipment is to generate ESD-like pulses. Once generated, these pulses are fed to a device under test (DUT) to evaluate the performance of the DUT operating under an ESD stress. Figure 1.8 shows schematically the concept of the ESD testing using a TLP tester. First, a transmission line with an impedance of 50 Ω is charged by a voltage supply. The discharging of the charged transmission line generates ESD-like pulses with various user-definable pulse amplitudes, rise times, and widths. A set of DUT's time-dependent voltage and current waveforms can then be measured under a particular pulsing stress. Taking the voltage and current values in a region where the voltage and current waveforms are fairly constant with respect to time yields a point in the quasi-static I–V plot. Repeating this process using pulses with different amplitudes results in many points in the quasi-static I–V plot, and connecting these points gives rise to the quasi-static I–V curve obtained from the TLP testing. This I–V curve provides useful information on the DUT's ESD performances, including the trigger voltage, holding voltage, on-state resistance, and failure current. The TLP tester can also measure the leakage current of the DUT after it is subject to a pulsing stress. The stress point at which the leakage current is increased significantly is defined as the failure point, and the corresponding current is the failure current. For HBM, the maximum protection capability of a DUT, in terms of the voltage, can be estimated by the product of the TLP failure current and the typical human body skin resistance of 1500 Ω. Such a simple correlation has been commonly used to benchmark the robustness of an ESD protection structure subject to the HBM stress. For the characterization of the DUT under the CDM condition, the very fast TLP tester that generates relatively faster and shorter pulses is used. For the system-level ESD event, such as that described by the IEC standard, the testing is typically carried out by the IEC gun air gap or contact testing.

REFERENCES

1. J. Vinson, G. Croft, J. Bernier, and J.J. Liou, *Electrostatic Discharge Analysis and Design Handbook*, 6 chapters, 300 pages, Kluwer Academic Publishers, Norwell, MA, December 2002.
2. J. Vinson and J.J. Liou, "Electrostatic discharge in semiconductor devices: protection techniques," *Proceedings of the IEEE*, vol. 88, pp. 1878–1900, December 2000.
3. M. Brandt and S. Halperin, "What does ESD really cost?" *Circuits Assembly Magazine*, June 1, 2003.
4. Z. Liu, J. Vinson, L. Lou, and J.J. Liou, "An improved bi-directional SCR structure for low-triggering ESD applications," *IEEE Electron Device Letters*, vol. 29, pp. 360–362, April 2008.
5. J.C. Lee, G.D. Croft, J.J. Liou, W.R. Young, and J. Bernier, "Modeling and measurement approaches for electrostatic discharge in semiconductor devices and ICs: an overview," *Microelectronics Reliability*, vol. 39, pp. 579–594, May 1999.
6. W. Liu, J.J. Liou, J. Chung, Y.H. Jeong, W.C. Chen, and H.C. Lin, "Electrostatic discharge (ESD) robustness of Si nanowire field-effect transistors," *IEEE Electron Device Letters*, vol. 30, pp. 969–971, September 2009.
7. Y. Li, J.J. Liou, J. Vinson, and L. Zhang, "Investigation of LOCOS- and polysilicon-bound diodes for robust electrostatic discharge (ESD) applications," *IEEE Transactions on Electron Devices*, vol. 57, pp. 814–819, April 2010.
8. A. Amerasekera and C. Duvvury, *ESD in Silicon Integrated Circuits*, John Wiley & Sons, New York, 2002.
9. S.H. Voldman, *ESD Circuits and Devices*, John Wiley & Sons, New York, 2006.
10. S.H. Voldman, *Latchup*, John Wiley & Sons, New York, 2007.
11. K. Van Mele, *Effective ESD Strategies in Nano-CMOS IC Design*, Sarnoff Europe, Gistel, Belgium, 2007. http://www.chipestimate.com/techtalk.php?d=2007-12-04.

2 Design of Component-Level On-Chip ESD Protection for Integrated Circuits

Charvaka Duvvury

CONTENTS

2.1 INTRODUCTION

An introduction to electrostatic discharge (ESD) protection has been described in Chapter 1. The purpose of this chapter is to present an overview of integrated circuit (IC) protection design methods that involve various clamps based on the protection needs to meet the design applications. More details of specific protection devices are discussed in Chapters 5, 6, 12, and 13.

IC chips need protection against ESD at all pins of the packaged device, and design methods have been well documented [1,2]. The ESD clamp is ideally in a high impedance state with tolerable capacitive load and triggers only when an ESD pulse is detected. This is conceptually shown in Figure 2.1 for protecting an input/output (I/O) circuit. With the occurrence of an ESD pulse on the IC pad, the protection device clamps a major portion of the ESD current energy to the ground bus. The clamp device must be fully compatible with the I/O function.

Component ESD protection first requires an understanding of what is being protected. In addition, the requirements of the ESD target levels for various models need to be established. The protection design without this information is only an *ad hoc* approach and may not meet the qualification requirements or may just end up being overdesigned. This type of overdesign can give a boost to the claims of competent ESD design, but it can also lead to higher than necessary capacitance and an area penalty with a large protection device.

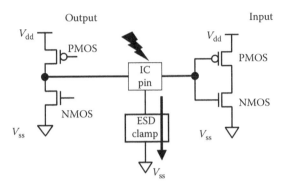

FIGURE 2.1 General ESD protection strategy.

In formulating the protection strategy, any type of intuitive design is undesirable as it is often guaranteed not to work consistently. An empirical method using data from test structures is a practical and safe approach. An approach based on simulations is gaining popularity although some test structure analysis up front is still required.

An overall systematic approach is to identify the design targets and formulate a strategy for the different types of pins being protected. These could be input, output, bidirectional, and power pins. Further, thinner gate-oxide inputs face restriction of the voltage rise at the gate from the discharge current through the ESD clamp's on-resistance and may force lowering the expected ESD levels. High-speed inputs and radio-frequency (RF) I/Os fall into these categories.

2.2 PROTECTION STRATEGY

The following should be the overall goals for protection designs:

- Clamp the ESD voltage to shunt the ESD stress current
- Turn on fast with <300 ps response
- Carry large currents of 1–2 A or more for 150 ns
- Have low on-resistance
- Occupy minimum area at the bond pad
- Impose minimum capacitance
- Introduce minimum series resistance
- Be immune to process drifts
- Be robust for numerous pulses
- Offer protection for the human body model (HBM) and the charged device model (CDM)
- Not interfere with the IC's functional testing
- Not cause increased IDDQ or I/O leakage
- Survive the burn-in tests
- Not cause latch-up or electrical overstress (EOS) failures

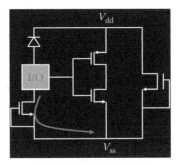

FIGURE 2.2   Local clamp approach. The ESD current is steered away from the I/O pad to V_{ss}.

Thus, the requirements for effective and efficient protection designs are beyond just meeting the ESD target levels. The protection device choice and the implementation with layout can also lead to some unexpected latch-up problems or even EOS in mixed-voltage designs [3].

There are essentially two types of protection design styles: (1) local clamp approach and (2) rail clamp approach. As shown in Figure 2.2, the local clamp approach involves direct ESD current path to ground (V_{ss}). Such a clamp must trigger quickly and carry enough ESD current before any voltage rise at the pad can damage the I/O buffer.

The local clamp usually involves a snapback device such as a gate-grounded NMOS (GGNMOS), which goes into parasitic npn bipolar conduction, as indicated in Figure 2.3. The device triggers at V_{t1}, goes into snapback at V_{sp}, and carries ESD current until it fails at (V_{t2}, I_{t2}). The failure current I_{t2} must be above the ESD current level for target spec level (e.g., 0.67 A for 1 kV HBM or 1.3 A for 2 kV HBM). The GGNMOS is a popular type of protection clamp, and there are different layout techniques that are followed to make it more efficient.

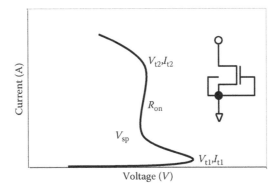

FIGURE 2.3   I–V behavior of a snapback device used for local clamp designs.

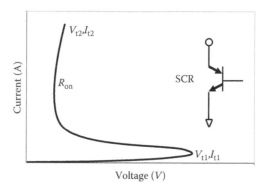

FIGURE 2.4 I–V behavior of an SCR device used for local clamp designs.

Instead of the snapback NMOS, it is also a common strategy to use a silicon-controlled rectifier (SCR) device for the local clamp. The current–voltage (I–V) behavior shown for the typical SCR in Figure 2.4 has similar trigger characteristics except that it holds at a lower voltage, has lower on-resistance, and as a result is much more efficient to carry the ESD current.

The local clamp approach can suffer from relatively higher capacitance that has some unwanted impact on the I/O circuit performance because the protection device has to be large enough to handle the necessary level of ESD current. This approach is widely used either with the diode to V_{dd}, as shown in Figure 2.2, or without the diode to V_{dd}. For fail-safe requirements, a diode to V_{dd} is not allowed by application engineers, and for these cases, the local clamp approach is the best option. Details of fail-safe protection are described in Chapter 12. Note that the total protection must always also involve a clamp placed between V_{dd} and V_{ss}.

The second, the more practical and currently increasingly popular, approach is shown in Figure 2.5, where the ESD current is diverted away from the pad toward the V_{dd} rail, forcing the V_{dd} clamp to carry the ESD current [4,5]. Note that the diode to V_{ss} at the I/O pad is in reverse bias and blocks the ESD current to ground while

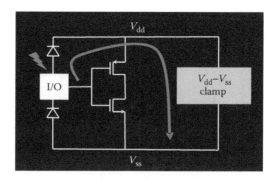

FIGURE 2.5 Rail clamp approach. ESD current is steered away from I/O pad to V_{ss} via the V_{dd} clamp.

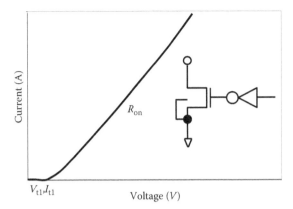

FIGURE 2.6 I–V behavior of an MOS device used for rail clamp designs.

the current goes through the upper diode to charge up the V_{dd} capacitance. Even in this case, the voltage at the I/O pad can eventually increase to affect the I/O buffer devices.

In its basic operation, the rail clamp approach involves the ESD current conduction through a large MOS device between V_{dd} and V_{ss}. The MOSFET uses gate control (see the inset in Figure 2.6) with an inverter circuit to keep it on for about 1 μs. There are numerous variations of the rail clamp design, which will be discussed in Chapter 5. As a MOS device is used for carrying the ESD current, there is no snapback involved, as shown in Figure 2.6.

The MOS device turns at V_{t1} and carries the ESD current until the voltage reaches the breakdown voltage limits of I/O buffers. That is, the device is not driven up to the point of failure, but the voltage buildup is usually the limiting point. In this rail clamp approach, the capacitance load is relatively small depending on the sizes of the diode to V_{ss} and diode to V_{dd}.

Another inherent advantage is that the NMOS also offers protection between V_{dd} and V_{ss}.

2.3 DESIGN WINDOW

To follow an appropriate protection strategy, the so-called design window must be first understood. This is shown in Figure 2.7 [6] as the "the ESD protection window." The turn-on of the protection device should not interfere with the IC operation, which forms the limit on the left side. Note that there are two protection device turn-on points depending on the design style. For the local clamp approach, the trigger voltage (just before snapback) defines this point, whereas for the rail clamp approach, the turn-on is at the V_{dd} diode conduction point. The right-hand-side border is defined by the gate-oxide breakdown voltage (for inputs), or it can also be defined by the transistor junction breakdown voltage (for outputs). Although the local clamp trigger might be close to the breakdown boundary, this is not very critical. This is because in most cases the clamp snaps back in voltage safely away from

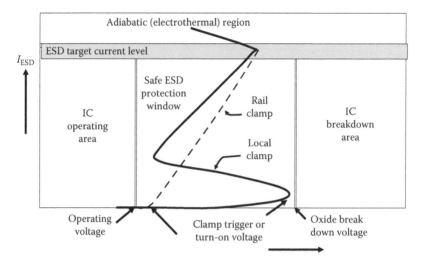

FIGURE 2.7   Basic ESD design window.

the breakdown limit within just a few hundred pico seconds. That is not sufficient time to cause damage. As the current is increased after trigger (per the local clamp approach) or beyond turn-on (per the rail clamp approach), the voltage with either design increases and reaches close to the breakdown regime. This voltage rise at the pad should not reach the breakdown points of the input gate oxide or the output drain junction, either of which forms the reliability limit. To meet the required ESD target level, the corresponding ESD current level must be safely reached. Note the "ESD target current level" in the figure. The protection device must not enter failure before this level is reached. For example, for a 2 kV HBM target with ESD regime oxide breakdown voltage (BV_{ox}) of 10 V, the voltage rise from 1.3 A of ESD current must be <10 V. Advanced technologies with thinner gate oxides pose a serious challenge as this design window narrows [7]. For technologies having BV_{ox} of <5 V, the 2 kV HBM target is a challenge. The current density through the metal leads of the protection device can also reach a limit, and this is shown as the adiabatic region in the figure. Thinner metal leads have higher resistance, and thus unless they are intentionally widened to compensate for this, they will offer more effective resistance to ESD current. That is, the voltage builds up to a higher value for the same ESD current magnitude. Widening the metal leads to meet the ESD current levels would increase the capacitance at the pad and reduces the possible speed performance. These issues would require a reexamination of the ESD target levels [8,9].

There are several different ESD clamp design approaches, but most of them are based on either a snapback device or a rail clamp method. In some cases, a large reverse-biased diode can also be used for protection, but this is mostly inefficient. Five different concepts with their typical clamping voltages and their respective on-resistance behavior are shown in Figure 2.8.

Consider first a 350 nm CMOS technology with a BV_{ox} of 10 V. Device 1 is a forward-biased diode, which is efficient to carry current more than necessary for

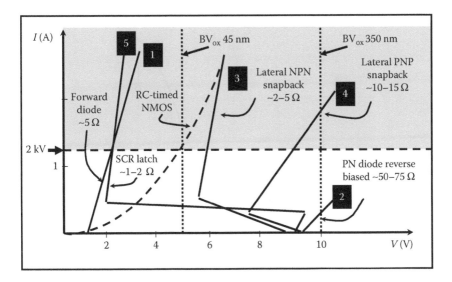

FIGURE 2.8   Design window with protection device concepts.

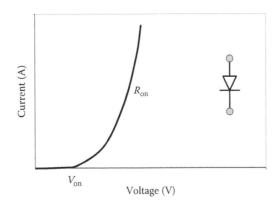

FIGURE 2.9   Characteristics of a forward-biased diode.

2 kV HBM target but is impractical because of leakage, unless it is used for 1 V V_{dd} applications. The typical I–V characteristics for a forward-biased diode are shown in Figure 2.9. The V_{on} is about ~0.5 V and the R_{on} is 100 Ω-μm, and thus a 200 μm perimeter diode will have an on-resistance of only 0.5 Ω and can carry 4–5 A. That is a figure of merit (FoM) of >20 mA/μm of ESD current before failure. The diode in forward mode is useful as a protection element in the rail clamp scheme of Figure 2.5.

Device 2 is a reverse-biased diode, which can break down below 10 V but has such high on-resistance that it cannot be effective to protect an input gate oxide with a BV of 10 V. The R_{on} under reverse mode is as high as 1 kΩ-μm. The typical characteristics for a reverse-biased diode are shown in Figure 2.10, where its V_{bd} is

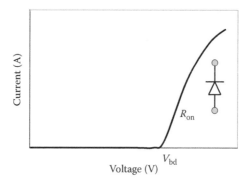

FIGURE 2.10   Characteristics of a reverse-biased diode.

often >>10 V, making it ineffective to protect a gate oxide or an output junction. For a 200 μm perimeter diode, it can only carry about 300 mA (FoM of ~1.5 mA/μm) before failure. A larger diode to meet 2 A of ESD current can be designed but is not practical. Instead, both forward and reverse diodes are effectively used as shown in Figure 2.5 to meet both positive and negative polarity protection. A more useful approach is to build two-stage diode network, which will be described later Figure 2.18.

Device 3 is an npn formed with an MOS device, which can provide >>2 kV HBM protection very efficiently. This is usually a GGNMOS or a gate-coupled NMOS (GCNMOS) (see Figure 2.11). These clamps can provide more than 2 A of ESD protection for a device width of 400 μm (FoM of >5 mA/μm). In a bipolar technology, this could be an npn device, which is even more efficient than the parasitic MOS npn.

Device 4 is a parasitic pnp device from a PMOS, which has a relatively higher breakdown voltage and higher on-resistance and is therefore marginally effective. Finally, device 5 is an SCR, which can trigger below 10 V with a very low clamping voltage and considerably low on-resistance. It can easily provide >4 kV HBM

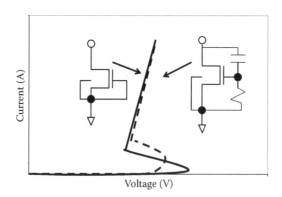

FIGURE 2.11   Gate-grounded NMOS (left, solid curve) versus gate-coupled NMOS (right, dashed curve).

protection for a device width of 100 μm (FoM of >40 mA/μm). However, a critical problem with the SCR is whether it can trigger fast enough (especially under CDM) and have low-magnitude overshoots to protect the gate oxide before clamping to low voltage. For this reason, often a two-stage protection is used in conjunction with an SCR to buffer any damage to the gate oxide. This is discussed in the next section.

Going back to Figure 2.5, if now an advanced technology of 45 nm is considered, the BV for the oxide is reduced to 5 V. This means the new protection devices for this advanced technology must clamp below this voltage. The trigger voltages for devices 3, 4, and 5 will also be inherently lower for the advanced technology nodes, but still the relatively higher amount of reduction in BV for the oxide makes it more difficult to achieve efficient protection. In contrast, device 4 with the rail clamp would still be effective, and 2 kV HBM protection can be practically achieved. It should be noted again that with device 4, a clamp between V_{dd} and V_{ss} that is compatible with the new advanced technology is needed. Even more advanced technologies with smaller feature sizes would place higher constraints on the ESD design on all the protection clamps discussed, and this issue is discussed in the final section.

2.4   PROTECTION DESIGN METHODOLOGY

When utilizing snapback protection designs, the concept of primary and secondary stages is important (see Figure 2.12). The primary device should shunt most of the ESD current while the purpose of the secondary element is to protect the input from gate-oxide damage or the output from transistor junction breakdown. In this scheme, the design of the isolation resistor is important as it serves two purposes: (1) limits the ESD current into the I/O devices, and (2) provides the voltage drop to trigger the primary. Without the isolation resistor, the scheme is not effective. For a useful strategy, the secondary must generally always trigger at a lower voltage than the primary. In some cases, the primary might trigger at a lower voltage than the secondary, but the primary clamp voltage will increase from there because of the $I \times R$ drop in the primary clamp impedance. This is why the resistor and secondary clamp are needed.

As shown in Figure 2.12, the primary devices can be snapback (NMOS, SCR) or dual diodes (Figure 2.5) to rail clamps between V_{dd} and V_{ss}. When diodes are used

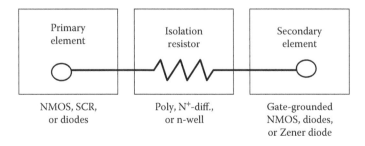

FIGURE 2.12   A typical protection scheme with primary and secondary devices.

for primary protection, they should not be operated in reverse mode where they would have high on-resistance and low ESD current failure (~1 mA/μm). On the other hand, a GGNMOS (see Figure 2.11) is effective as long as it is sized properly to carry the target ESD current level. An SCR device tends to have higher trigger voltage, and the design of the isolation resistor becomes more critical in this case [10]. Any combination of devices for the primary and secondary can be used as long as they are compatible for trigger and ESD operation.

The isolation resistors are optimized based on design conditions. A diffusion resistor (made of N$^+$) is optimal because its effective value increases during the high-current ESD. Its value is typically 25 Ω/sq. without silicide. One limitation is that its breakdown to the P-substrate is only about 15 V, and hence the high end of the resistor cannot support much voltage. Even more effective is the n-well diffusion resistor due to lower doping of the well diffusion. It can not only support a higher voltage but in the pinch-off regime can also offer resistance that is >5X from its low current value. This behavior is shown in Figure 2.13. The N$^+$ stripes in the n-well define the resistance, $R = L/W \times \rho$, where ρ is the resistivity of the diffused well region, typically 600 Ω/sq. for the n-well. The I–V curve shows that the resistance in the pinch-off regime increases until the resistor itself snaps back at the avalanche voltage, V_{av}. For this reason, the primary device such as the SCR must trigger below this level to safely use the resistor as an isolation element. The resistor can be ideally used in both the linear and saturation regions while avoiding the breakdown point at V_{av}. The spacing L critically controls the V_{av} point.

The problem with these diffusion resistors is that they have higher capacitance and some associated leakage. Poly resistors on the other hand are preferred by analog designers because they have lower capacitance, are more stable, and have better linearity under voltage bias. The typical I–V curve for a poly resistor is shown in Figure 2.14. The poly resistors are isolated from the substrate heat sink and hence can be damaged if excess current is passed through them before the primary device triggers [2]. They should operate only in the initial region before bending of the curve, which indicates early heating shown in the figure. One way to make them effective is to design them to be wider. That is, to be able to drive a larger amount of current before damage. For this reason, they have to be laid out both wider and longer to obtain the same resistance value. The resistance is $R = L/W \times \rho$, where ρ is polysilicon resistivity. The poly can also be silicided. For both diffusion and poly resistors, they have to be first characterized for their current-carrying capability and then design them in the protection scheme according to Figure 2.12, such

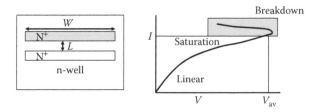

FIGURE 2.13   Characteristics of a diffusion resistor.

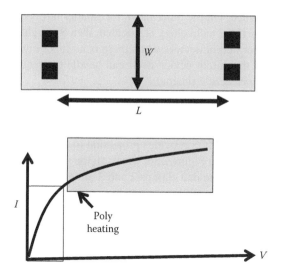

FIGURE 2.14   Characteristics of a poly resistor.

that only a safe amount of current is allowed through them before the primary triggers. This allowed current can easily be estimated through prior test structure characterization.

2.5   OVERALL PROTECTION STRATEGY

Following the two-stage protection strategy described in Figure 2.12, the I/O buffers are commonly protected, as shown in Figure 2.15. Here clamp 1 is the primary device and clamp 2 is the secondary device. It can be seen that the series resistor (R_1) is placed only for the input side, whereas for the output side the pad connection is directly made to the output diffusion with an optional resistor (R_2). R_2 depends on the size of output devices and specific technology features. For large buffers such as 4 and 8 mA, the output NMOS may be able to handle enough of the ESD current that

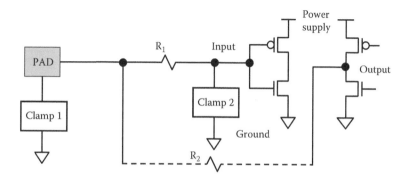

FIGURE 2.15   I/O protection scheme.

the resistor is not needed. On the other hand, for smaller buffers (1 or 2 mA), R_2 is usually needed. Also, if the technology is silicided, then R_2 might be needed for any buffer device size. For output devices when there is no input involved, the primary can be eliminated if the output device itself can handle the required ESD current, making it into a *self-protection* scheme.

One issue with the design scheme of Figure 2.15 is that there is no protection for the input PMOS gate oxide. As a result, there is some vulnerability to the CDM stress. Note that for protection on the output side clamp 3 is not needed because the inherent diode in the PMOS provides protection to V_{dd}. A more comprehensive scheme for inputs is shown in Figure 2.16, where all the necessary clamps are placed for the input buffer. Also note that clamps 2 and 3 are placed close to the buffer for the best achievable protection while avoiding bus resistance effects such as from R_{ss} in Figure 2.16 [11].

For many CMOS technology applications, the dual diode is a more popular scheme because it is simple and straigthforward in implementation. Figure 2.17 shows a multiple I/O protection scheme, all with same-sized diode clamps and sharing a common V_{dd} cell that usually involves a MOS rail clamp. Note that an additionla substrate diode in the V_{dd} cell is also important to cover all positive and negative current paths during ESD testing.

Although each I/O ESD cell with dual diodes is the basic approach, additional secondary clamps are important for each cell to ensure good CDM protection. This is shown in Figure 2.18, where D_1 and D_2 are the primary clamps (typically 200 μm in perimeter) while D_3 and D_4 are the smaller secondary clamps (typically 50 μm in perimeter). The resistor R is important and can be about 50 Ω. When CDM stress is applied at the I/O pad, the current through D_3 or D_4 depending on the polarity of stress will provide the voltage drop margin to effectively protect the input gate oxide. For this reason, the combination of R and D_3/D_4 is often called the "CDM clamp."

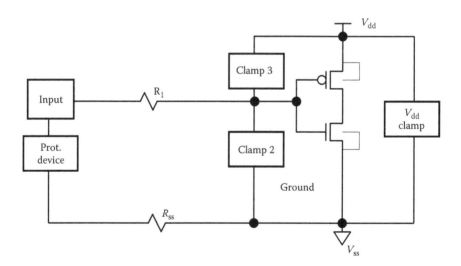

FIGURE 2.16 HBM/CDM protection scheme for input.

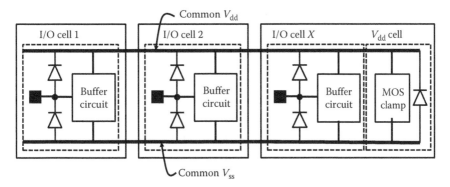

FIGURE 2.17   A common approach for multiple I/O protection with dual diodes.

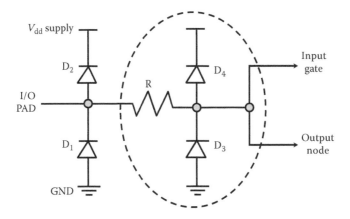

FIGURE 2.18   Advanced HBM/CDM protection with dual diodes.

The value of R plays a significant role in limiting the I/O buffer speed [12]. For high-speed serial (HSS) link designs with >20 Gbits/s, the R values may have to be only a few ohms. Some of this will reduce the achievable CDM protection for microprocessors in large IC packages [12].

The overall protection must also consider the various current paths for HBM or CDM testing requirements. Consider the multiple-domain mixed-voltage IC design in Figure 2.19.

The periphery could be at one voltage level (1.8, 3.3, or 5 V) while the core with high-density circuits could be at a different lower voltage level (1.2, 1.5, or 3.3 V). In this case, both the V_{dd} and V_{ss} connections are different. I/Os in the periphery are connected to I/OV_{dd} and I/OV_{ss}; core circuits are connected to CoreV_{dd} and CoreV_{ss}. Both the periphery and the core require their own resepective compatible clamps (clamps 3 and 4) with parallel substrate diodes in each case. Clamp 6 most commonly consists of antiparallel diodes between I/OV_{ss} and CoreV_{ss}. While testing the combination of I/O to CoreV_{ss}, clamp 1 along with clamp 6 is essential. For I/O to

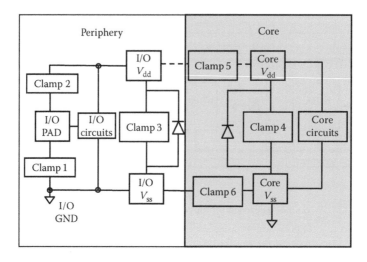

FIGURE 2.19 Mixed-voltage domain protection.

CoreV_{dd}, clamp 1 plus clamp 6 plus diode parallel to clamp 4 will take care of the current path. But alternately, clamp 2 plus clamp 5 (also consisting of antiparallel diodes) can be designed in for this path. However, care must be taken not to accidentally forward bias these diodes between the differnt power supplies during power sequencing. For these reasons, clamp 5 is usually not used.

In the same manner, system-on-chip (SoC) designs would have analog and digital domains, and these also need a careful protection strategy. For example, for the SoC shown in Figure 2.20, small diodes are needed at the interface to protect for different combination stress tests for HBM or for overall CDM protection. For die-to-die interfaces, a similar strategy is important when I/O from die 1 interfaces with I/O from die 2.

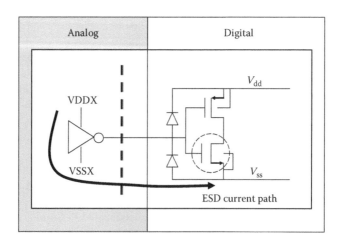

FIGURE 2.20 Multiple SoC domain protection.

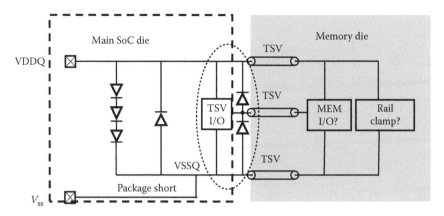

FIGURE 2.21   Three-dimensional IC protection for CDM. (Data from Schulmeyer, K. et al., "ESD considerations for 3D ICs," in *Semicon West*, 2012.)

This will be especially important for three-dimensional IC designs, as shown in Figure 2.21. Here the memory IC is interfacing with the SoC IC with through-silicon via (TSV) technology. Small diodes are again placed at the inteface to protect for CDM [13].

2.6   SUMMARY AND FUTURE DIRECTIONS

As was discussed in this chapter, the on-chip ESD protection strategy involves many facets. The most important requirements are the protection designs that do not interfere with the I/O functions while providing the necessary protection for HBM and CDM qualification requirements. The total protection involves in addition to the peripheral I/O pins the internal chip protection between the power supplies.

There are different styles of effective protection clamps that are commonly used, and the choice depends on the technology advances and the design constraints. These constraints are defined by the ESD design window, which must ensure that the I/O buffers are not damaged while meeting the ESD targets. The protection designs must also consider the IC chip applications such as mixed voltages and/or multiple power domains.

As technologies advance and higher-speed circuits are more in demand, there are many new restrictions on the ESD design window. In Chapter 13, more details of the advanced technologies and the impact on ESD are discussed. In general, the I/O designers should not be burdened with excessive ESD target levels in such cases, but instead the quality groups must negotiate with the customers to focus on more realistic but safe target levels [8–9]. For example, as shown in Figure 2.22, for 10–20 Gbits/s high-speed applications in microprocessors, the ESD design window may be restricted to <1 kV HBM for 32 nm technologies and beyond [12]. Similarly for microprocessors built in large IC packages, the CDM levels are reduced to 250 V or even lower, as shown in Figure 2.23 [9].

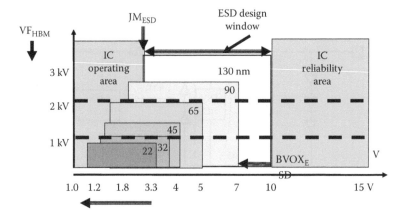

FIGURE 2.22 The evolving ESD design window. (Data from Duvvury, C., "Silicon technology impact on ESD qualification," in *VLSI Design Conference*, Pune, India, January 2013.)

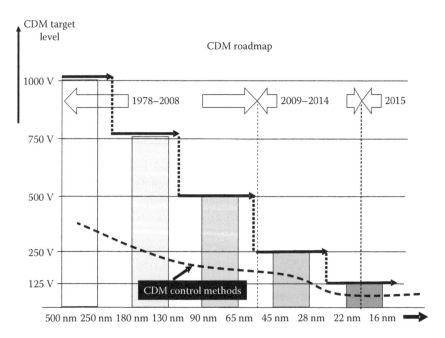

FIGURE 2.23 The evolving roadmap for CDM. (Data from White Paper 2, *A Case for Lowering Component Level CDM ESD Specifications and Requirements*, http://www.esda.org/documents/IndustryCouncilWhitePaper2.pdf, alternately known as JEP157.)

ACKNOWLEDGMENT

Grateful thanks are offered to Steve Marum (Texas Instruments, retired) for his helpful comments and suggestions.

REFERENCES

1. T. Maloney and S. Dabral, *Basic ESD and I/O Design*, John Wiley & Sons, New York, 1998.
2. E.A. Amerasekera and C. Duvvury, *ESD in Silicon Integrated Circuits*, 2nd Edition, John Wiley & Sons, New York, 2002.
3. J. Salcedo-Suner, R. Cline, C. Duvvury, A. Cadena-Hernandez, L. Ting, and J. Schichl, "A new I/O signal latchup phenomenon in voltage tolerant ESD protection circuits g," in *Proceedings of the Reliability Physics Symposium Proceedings*, 2003. 41st Annual IEEE International, pp. 85–91.
4. R. Merrill and E. Issaq, "ESD design methodology," in *Proceedings of 15th EOS/ESD Symposium*, 1993, pp. 233–237.
5. E.R. Worley, R. Gupta, B. Jones, R. Kjar, C. Nguyen, and M. Tennyson, "Sub-micron chip ESD protection schemes which avoid avalanching junctions," in *Proceedings of 17th EOS/ESD Symposium*, 1995, pp. 13–20.
6. G. Boselli, J. Rodriguez, C. Duvvury, and J. Smith, "Analysis of ESD protection components in 65 nm CMOS technology: Scaling perspective and impact on ESD design window," in *Proceedings of 27th EOS/ESD Symposium*, 2005.
7. C. Duvvury, "ESD qualification changes for 45 nm and beyond," in *IEDM Tech. Digest*, 2008, pp. 1–4.
8. White Paper 1, *A Case for Lowering Component Level HBM/MM ESD Specifications and Requirements*, http://www.esda.org/documents/WhitePaper1_HBM_MM_2010.pdf, alternately known as JEP155.
9. White Paper 2, *A Case for Lowering Component Level CDM ESD Specifications and Requirements*, http://www.esda.org/documents/IndustryCouncilWhitePaper2.pdf, alternately known as JEP157.
10. C. Duvvury and R. Rountree, "A synthesis of ESD input protection scheme," in *EOS/ESD Symposium Proceedings*, 1991, pp. 88–97.
11. T. Maloney, "Designing MOS inputs and outputs to avoid oxide failure in the charged device model," in *EOS/ESD Symposium Proceedings*, 1988, pp. 220–227.
12. C. Duvvury, "Silicon technology impact on ESD qualification," in *VLSI Design Conference*, Pune, India, January 2013.
13. K. Schulmeyer, C. Duvvury, and R. Cline, "ESD considerations for 3D ICs," in *Semicon West*, Sematech ESD Workshop, San Francisco, CA, 2012.

3 ESD and EOS: Failure Mechanisms and Reliability

Nathaniel Peachey and Kevin Mello

CONTENTS

> Amid the turmoil of battle, there may be seeming disorder and yet no real disorder at all.
>
> **Sun Tzu in *The Art of War***

3.1 INTRODUCTION

The solution of a problem can often be clarified by ordering it into rational classifications or categories. This is particularly applicable to the understanding and analysis of the entire scope of electrostatic discharge (ESD) and electrical overstress (EOS) failures. When analyzing a failure that is presumably an ESD or an EOS failure, deciphering the failure mechanism is often the first step in addressing the problem. This chapter will focus on a systematic presentation of the failure mechanisms and related reliability issues associated with ESD and EOS events. In the most basic definition, EOS is simply an electrical stress that is applied to a device beyond that for which it has been designed. This may or may not be transient or dynamic in nature. By this definition, EOS will result in damage of the device. Indeed, some would define EOS as the damage that results from an electrical stress. ESD, on the other hand, is an electrical stress that results from the transient discharge of electrostatic energy. ESD may or may not result in failure of the device because devices are normally intended to tolerate some level of ESD stress. In this chapter, we will focus more directly on ESD but will also discuss some implications of failures resulting from other electrical stresses.

In semiconductors, the elements that can be affected by an ESD stress are the following [1]:

1. *Oxides*. These are gate oxides, capacitor dielectrics, and any other oxide structure.
2. *Junctions*. These include the transistor and bipolar junctions as well as diodes.
3. *Metals/materials*. In addition to back end of line (BEOL) metals, these include silicides, resistors, and any material that electrically connects semiconductor elements.

The most common failure mechanisms in oxides have to do with oxide rupture or the generation of defects within the oxide through hot carriers. At the core of these hot carrier failures are energetic electrons that cause damage to the oxide. Within junctions, failure is associated with thermal processes. Heat drives the rearrangement of the semiconductor structure and composition that fundamentally alters the junction characteristics. Heating can result in either surface breakdown associated with the junction or internal breakdown at the junction interface in the bulk of the material [2]. Metals and materials failure is also due to thermal processes but also include transport of the material itself. This chapter focuses on these three failure categories as well as the failure mechanisms and reliability issues associated with each.

Much of what is understood about failure mechanisms in semiconductor devices has been reported decades ago. Early work in the field of thermal failure of semiconductors due to electromagnetic pulses in nuclear science prepared the way for an understanding of semiconductor failure mechanisms under ESD and EOS stresses. However, although the fundamental physics of ESD failure mechanisms is well understood, the analysis of particular failures in a complex semiconductor circuit remains a challenging task. Thus, this challenge is the focus of this chapter.

3.2   OXIDE FAILURE MECHANISMS AND RELIABILITY

Oxide failures are generally associated with a charged device model (CDM) event. With the advent of automated assembly of electronic boards, handling of devices by personnel in the factory is no longer a major threat in the modern semiconductor factory. Instead, most ESD failures result from either devices becoming charged in a stray field and then being discharged when they come in contact with a grounded surface or a device contacting a charged surface directly. Current state-of-the-art factory control standards such as ANSI/ESD S20.20 require conductive surfaces to be grounded so that in a factory that is properly controlled, the chances for a device to be damaged by contacting charged, isolated conductors are minimized. Nevertheless, devices continue to face the threat of becoming charged during processing and then being discharged to a grounded conductor. Damaging CDM events most generally are associated with oxide failures within the device. Thus, understanding the oxide failure mechanisms and the latent reliability issues associated with ESD stresses on oxides continues to be a concern.

Earlier investigations of gate-oxide damage mostly addressed the injection of hot electrons during source to drain current stressing, avalanche currents, and snapback [3–5]. More recently, attention has been focused on the effects of a CDM-type event on the gate oxide directly. This attention has increased as both the feature sizes and oxide thickness have decreased in advanced technologies [6–8]. Indeed, CDM concerns have come to dominate the handling of ESD-sensitive devices in modern factories and assembly plants.

3.2.1 LATENT EFFECTS

The most critical structure when it comes to gate-oxide reliability is the n-FET device in inversion [3,9]. Indeed, the time-dependent dielectric breakdown (TDDB) ratio between p-FETs and n-FETs has been found to be approximately 1.25 regardless of oxide thickness [7]. Consequently, the n-FET device in inversion has been the focus of much of the study of gate-oxide failure and reliability. Although junction failures follow slightly different time-to-breakdown characteristics in various short pulse regimes, the same is not the case for oxides. The power-law model typically used to describe time-to-breakdown appears to work equally well under DC or ESD time domains [7–9]. Following the power-law model, TDDB can be described as [8]:

$$t_{bd} = aV_G^{-n} \tag{3.1}$$

The exponent, $-n$, is an oxide thickness–independent factor. This factor can take on one of two values. Below approximately 5–6 V, this value has been measured as between 44 [8] and 48 [9] and above, and it becomes 30 for the n-FET in inversion [9]. The change of factors has been attributed to the shift from a Fowler–Nordheim tunneling mechanism at low voltage to a direct tunneling mechanism at higher voltage [10]. Figure 3.1 shows this factor difference as a "kink" in the $t_{63\%}$ time to failure versus voltage plot [9]. Fowler–Nordheim tunneling shows an accelerated failure versus voltage relationship for the lower-voltage regime. This, of course, assumes that the applied voltage remains below the actual failure voltage where oxide rupture occurs.

The thickness of the oxide is the most critical factor in determining the relative TDDB. For each nanometer of oxide thickness, there is approximately six decades of difference in the time to failure [7].

One of the more troublesome aspects of gate oxides is that they can sustain latent damage from ESD-type events even if the oxide does not fail immediately [6,9]. This can result in shortened oxide lifetimes, shifts in the threshold voltage, and increased leakage. The mechanism for this damage is the injection of hot carriers into the oxide that become permanently trapped in the oxide. To quantify the impact of the hot carrier

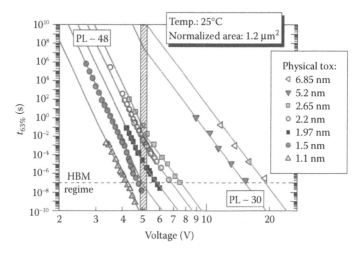

FIGURE 3.1 Voltage acceleration of n-FET stressed in inversion for seven oxide thicknesses. $t_{63\%}$ is derived from the Weibull plot, which is the cumulative failure level where 63% of the samples have failed.

injection, stress-induced leakage current (SILC) is measured for stressed and unstressed oxides. The effects of stress of the oxide become apparent in the subthreshold current–voltage (I–V) characterization region. Figure 3.2 shows the subthreshold region and the leakage shifts due to SILC. Fresh and stressed samples are compared along with the breakdown using the power-law model in Equation 3.1 [9].

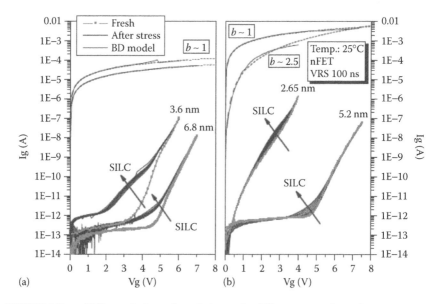

FIGURE 3.2 n-FET gate leakage degradation under 100 ns rectangular voltage ramp stress pulses for (a) 3.6 and 6.8 nm gate-oxide thicknesses and (b) 2.65 and 5.2 nm gate-oxide thicknesses.

Studies on oxides stressed to just below the breakdown voltage have been done to characterize the effects of ESD stresses. Latent effects in oxides of thickness greater than approximately 2.5 nm differ from those that are thinner [9]. For thicker oxides that are stressed to within 70% or 80% of their breakdown voltage, hot carriers become trapped in the oxide, degrading their performance. This degradation can be observed by increased SILC. Traps can be generated either as interface states or as states within the bulk of the oxide. Subsequent to stressing, the oxides will partially relax even at room temperature. However, although the oxide qualities improve, for thick oxides, the film does not completely recover. This is because of the nature of the traps. While interface traps can recover at room temperature, bulk states cannot or will only do so very slowly. Thus, the effect of the bulk traps is to change the long-term characteristics of the oxide. Oxides thinner than 2.5 nm that are stressed even up to 90% of their breakdown voltage do not show the same long-term SILC degradation. This results from the fact that these thin oxides do not have a "bulk" in which traps can develop and thus only have interface states.

For both thin and thick oxides, however, ESD stresses at 90% of their breakdown voltage or above does affect device reliability [9]. Even for thin oxides where no degradation was observed, overall dielectric lifetime was impacted. Based on these studies, a safety margin of approximately 10% to 15% below the actual breakdown voltage should be maintained when designing ESD protections for these oxides.

3.2.2 OXIDE FAILURE

Gate-oxide failure is generally due to one of two general mechanisms. First, avalanche current resulting from high source to drain voltage can generate hot carriers that can damage the oxide [4]. This will typically be observed in increased SILC, and breakdown is defined as a 10%–15% increase in leakage. The second mechanism involves direct stressing of the oxide across the gate in a CDM-like event. Breakdown resulting from stress directly across the gate oxide also begins with the increased leakage due to SILC. This is observed in a continuous shift of the gate leakage with continued or repeated stress. However, once breakdown or rupture occurs, there is an immediate increase in leakage current by several orders of magnitude. When the damage sites are analyzed, they are found to contain silicon in either amorphous or polysilicon form [11].

Although the result of a CDM type of failure is oxide failure or rupture, the mechanisms that cause this can be varied. The typical failure results from charging of internal capacitance followed by a very rapid discharge. The voltage increase across the oxide can build up very quickly and exceed the oxide breakdown voltage.

The vulnerability of a product to CDM failures is related to the inherent capacitance in the product. Much of this may be parasitic capacitance associated with either the design itself or the packaging used for the part. Radio frequency ports that incorporate capacitive tuning elements can be particularly vulnerable because these typically cannot tolerate the performance degradation associated with ESD protection circuitry. Failure of such a tuning capacitor can be observed in Figure 3.3. Other applications such as high-speed digital pins are also very sensitive to the additional parasitic capacitance

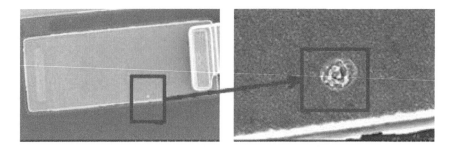

FIGURE 3.3   CDM failure showing the resulting oxide rupture.

of the ESD protection circuits. Other, more novel CDM failures have been reported. A recent report describes how long metal lines that are colinear and in close proximity can induce charge in each other [12]. In this case, charge that was generated in one metal line coupled into the adjacent line and caused the failure of an internal gate.

3.3   FAILURES IN SEMICONDUCTOR JUNCTIONS

One of the foundational building blocks of solid-state devices is the semiconductor junction and resulting diode. These provide the basic elements for the variety of transistors realized in semiconductor technologies. Fundamentally, the processes that cause damage to semiconductor junctions are thermal processes [2,13]. At the most basic level, failure in semiconductor junctions is largely because of heating of the material to near or above the melting temperature. Once this occurs, the material properties are permanently changed, and irreversible damage occurs. Another characteristic of the semiconductor junction is that it is not nearly as susceptible to latent damage as are oxides [14]. Unless the thermal energy causes a change of the crystal structure of the semiconductor, a redistribution of dopants, or a change in the alloy of the material, there is little chance for permanent damage or failure of the device [13,15]. ESD is a transient electrical stress that, in general, causes local hot spots where damage occurs. EOSs of longer duration tend to cause much more extensive damage in a device.

The mechanisms for thermal breakdown in junctions due to pulses were first reported over 30 years ago by various researchers, many of who were doing work for the Department of Defense or other governmental agencies [2,16–18]. At a fundamental level, thermal behavior in a material is a consequence of two competing mechanisms [13]. The first is the energy resulting from the pulse that impinges on the device, causing the temperature to rise. The second competing process is thermal diffusion bringing the temperature back toward ambient. The energy delivered in the pulse will determine the time to failure, and subsequently, the role that thermal diffusion plays in the power-to-failure threshold. In the shortest times to failure, the failure point can be considered to be a spherical defect. This is the approach taken by Tasca [16] and others [19]. In this case, the power to failure can be written as [20,21]:

$$P_f = \left(\frac{\rho C_p}{t_f} V \right) \Delta T \tag{3.2}$$

where:
ρ is density of the material
C_p is specific heat
V is volume of the spherical hot spot
t_f is time to failure
T is temperature

When the failure pulse transient is considerably shorter than the thermal diffusion time, the geometry of the hot spot does not contribute appreciably to the power-to-failure equation. Under these adiabatic conditions, Equation 3.2 is a very good approximation. When the failure pulse transient is slightly longer in duration such that the thermal diffusion length approaches the size of the junction, then the area of the junction interface must be considered. In this case, Equation 3.2 no longer describes the power to failure as well. The power to failure is then better described by Wunsch and Bell as [2]:

$$\frac{P_f}{A} = \sqrt{\frac{\pi \kappa \rho C_p}{t_f}} \Delta T \tag{3.3}$$

where:
κ is the thermal conductivity
A is the junction area

The pulse lengths for which the authors validated this equation were from 100 ns to 20 μs. Dwyer, Fanklin, and Campbell [13] relaxed the pulse length further and found that under even longer pulses, the thermal diffusion considerations are indeed three-dimensional. In the case of the diode, not only the junction area but also the depletion region will impact the power to failure. The equation that best describes the power-to-failure relationship is [22]:

$$P_f = \frac{4\pi\kappa a}{\ln\left(t/t_b\right) + 2 - \left(c/b\right)} \Delta T \tag{3.4}$$

where:
$t_b = b^2/4\pi D$
a, b, and c are the lengths of the three-dimensional heat dissipation region

When the time of the pulse becomes much longer than the diffusion time in the longest dimension of the heat dissipation region, the power to failure essentially becomes a constant that can be described by:

$$P_f = \frac{2\pi\kappa a}{\ln\left(a/b\right) + 2 - \left(c/2b\right)} \Delta T \tag{3.5}$$

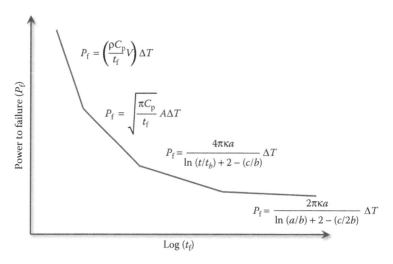

FIGURE 3.4 Power to failure versus log of time to failure.

The diffusion time–independent power-to-failure equation best characterizes the EOS failure mechanism. This power to failure versus time relationship can be shown graphically, as is done in Figure 3.4 [20,23].

There are several practical considerations that can be drawn from the failure mechanisms discussed earlier. The first rather obvious conclusion is that the pulse width greatly impacts the actual power required to cause failure. ESD pulses, while fairly high in instantaneous current, are quite short and rather local events. Thus, on-chip protection for some level of threat is practical because the ESD protection circuit itself can survive high-energy pulses that have a very short duration. However, for the longer EOS pulses or currents, it is very difficult to provide any significant on-chip protection given that the power to failure is proportionally much lower and thermal diffusion of the energy can be rather large.

Another consideration is that as the device geometries shrink, the thermal diffusion length reach the junction geometic sizes at much shorter pulse widths. As the power to failure is a function of thermal diffusion length and reduces as the junction dimensions shrink, the entire power to failure versus pulse width curve will be shifted lower and to shorter pulse widths. In practice, this contributes to the shrinking "safe-operating voltage" for ESD protection devices in deep submicron technologies.

3.3.1 FORWARD-BIASED BREAKDOWN

Considerably more research has been done for reverse-biased breakdown than for forward-biased breakdown. This is understandable because most failures encountered, particularly for earlier investigators, resulted from the failure of reverse-biased junctions. With the advent of rail-based ESD protection and particularly the active clamp, failures resulting from forward-biased junctions have become more common.

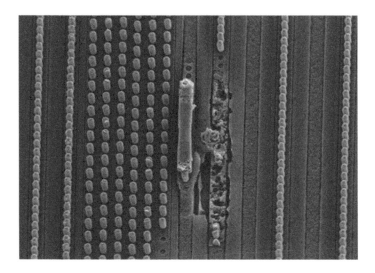

FIGURE 3.5   Electrical damage to an ESD protection diode due to forward-biased current.

An example of this is shown in Figure 3.5 where the ESD protection diode was overwhelmed with a forward-biased current and failed catastrophically.

In the forward-biased configuration, the voltage drop across the diode is merely the built-in potential barrier between the two regions in the small signal model [24]. Several earlier researchers have mentioned forward-bias failure modes but treated the forward-biased junction as having little or no voltage drop across the junction [13,16]. However, what was not considered is that a junction under high-current conditions will indeed have a significant voltage drop across the junction. This is mainly because of the on-resistance of the diode. A typical I–V plot of a diode is shown in Figure 3.6. In this example, at the peak current of a 2000 V HBM pulse (1.3 A), the voltage across the diode is nearly 3 V. The built-in potential for a forward-biased silicon diode is 0.7 V [24]. The remaining 2 V potential drop across the diode is then due to the on-resistance of the diode itself.

There are additional considerations that affect the failure mechanisms of forward-biased junctions in diodes and transistors. Under very fast transients, the current flow within the diode can be nonuniform and can result in a significant voltage overshoot [25]. The maximum voltage overshoot can be expressed as:

$$V_{\max} \propto \sqrt{\frac{J_{\mathrm{m}}}{\mu_{\mathrm{n}}\mu_{\mathrm{h}}N_{\mathrm{d}}t_{\mathrm{r}}}} \qquad (3.6)$$

where:
   J_{m} is current
   μ_{n} and μ_{h} are electron and hole mobilities
   N_{d} is donor concentration
   t_{r} is the pulse rise time

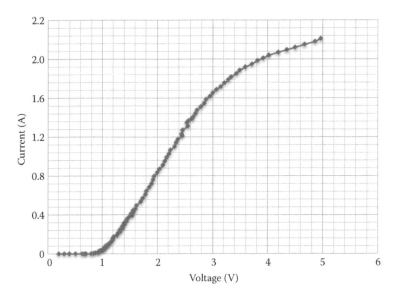

FIGURE 3.6 Transmission line pulse (TLP) results for a diode in forward-biased conditions.

Thus, the magnitude of the voltage overshoot is inversely proportional to doping concentration and electron mobility. In the cases of lightly doped diodes that are sometimes used for low parasitic capacitance applications, this voltage overshoot can be sufficiently significant such that the failure voltage is reduced at very fast rise times [25]. These authors also report filamentation in failed forward-biased diodes, a feature normally associated with reverse-biased junction failure.

3.3.2 REVERSE-BIASED BREAKDOWN

The more widely studied breakdown is that of the reverse-biased junction. Prior to thermal failure, the junction undergoes what has been defined by many researchers as *second breakdown* [15,16,18]. When a junction is reverse biased beyond the junction breakdown voltage, a current begins to flow that is generated by avalanche breakdown. However, as the voltage across the junction is increased even further, a second breakdown occurs as it heats up to a critical temperature. At this point, the temperature of the junction increases rapidly, changing the characteristics of the diode and causing the voltage across the junction to drop. This is the point at which thermal failure will then occur [17]. Second breakdown first occurs at a local hot spot but then can rapidly expand depending on the pulse width of the energy applied. In this manner, the reverse-biased breakdown behavior is described by the power-to-failure relationship in Figure 3.4. The second breakdown voltage is dependent on parameters such as the space charge width of the junction, which is dictated by doping levels [22]. Increasing the space charge width by decreasing carrier concentration will increase both the breakdown voltage and the second breakdown voltage. Defects can also impact the second breakdown as they provide sites for the formation of local hot spots [15,16].

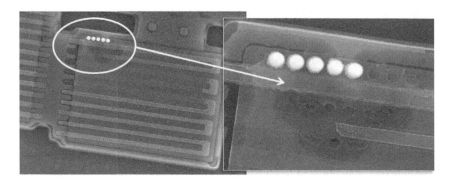

FIGURE 3.7 NMOS failure caused by an ESD pulse.

Semiconductor junction diodes are the foundation of the various transistor devices that comprise the ESD-sensitive circuits. Indeed, other than gate oxide–related failures, many ESD failure mechanisms for transistors can be understood from what has been learned about junction breakdown. However, in addition to the junctions themselves, the CMOS transistors have parasitic bipolar devices associated with them. It is the interplay of the parasitic bipolar devices with the ESD energy that results in snapback, which leads to one of the most common ESD-induced failures in transistors. Figure 3.7 shows an NMOS output driver that was damaged because of inadequate protection of the output pin.

NMOS devices are typically more vulnerable to junction failure than PMOS transistors. This is because of the fact that the parasitic npn bipolar device of the NMOS transistor has higher gain than the pnp parasitic device of the PMOS [26]. The increased gain of the NMOS parasitic device results in higher current and thus increased self-heating during an ESD event. This is also why many PMOS transistors display a reduced snapback if they have one at all.

3.3.3 Surface and Internal Breakdown

Junction breakdowns can be divided into either surface or internal breakdown [2,22,27]. Some of the earlier work on surfaces and their interaction with junction breakdown was done by Davies and Gentry [27]. Their work helped explain the observation that the P–N junction surface influenced the behavior of junction devices in breakdown and thermal failure. The electric field at the surface of the P–N junction is impacted by the contour of the surface. Furthermore, there can be a narrowing of the space charge width close to the surface inducing surface breakdown through localized avalanche breakdown current. Surface defects and anomalies can act to induce local hot spots, leading to early thermal failure.

Where surface breakdown becomes important is in the fully silicided lightly doped drain (LDD) transistor. As this device is the foundation for most modern semiconductor circuits, understanding the failures associated with surface breakdown is imperative. Amerasekera et al. studied and modeled the second breakdown and subsequent ESD robustness of these structures [22]. In this study, current flow was monitored using light emission spectroscopy. In a multi-fingered, fully silicided transistor, it was observed that not all fingers carried the current uniformly. Next, as the transistor was pushed toward snapback, only a single junction went into snapback. Finally, when the actual second breakdown was reached, only a single point along the junction in snapback emitted light and all of the rest of the device ceased to show current conduction. Typically, this results in filamentation at the point of failure. The emission experiment also underscores the observation that transistor breakdown and failure cannot always be predicted using the relationships described by Equations 3.2 through 3.4. Increasing the transistor size and thus the dimensions of the junction may not lead to an increase in the power to failure. Another observation made by these researchers was that the LDD device failed at a significantly lower power than the same device without the LDD implant. They concluded that the LDD device had a much shallower current injection region leading to decreased heat dissipation.

To overcome the lower power to failure that is exhibited by the silicided LDD device, a spreading or ballasting resistance can be added to the drain of the transistor [22,28]. Indeed, one of the most widely used ESD protection device is the NMOS transistor with the extended, silicide-blocked, drain. With the properly designed ballasting resistance, the NMOS ESD protection transistor does scale linearly with power to failure as is expected by the power-to-failure equations. The silicide-blocked length of the extended drain is typically greater than the thickness of the drain implant. This, in effect, preferentially favors bulk current such that the avalanche current necessary to turn on the parasitic device is generated in the bulk rather than at the semiconductor surface.

3.4   EOS FAILURE

Much has been written with respect to ESD failure mechanisms and reliability issues associated with these. However, the broader spectrum of EOS failures has received much less attention. This may be partially because of the diversity of threats that can result in EOS damage of devices. Over the past decades, ESD threats have been mitigated by on-chip ESD protection circuits and improved design practices. This has been coupled with much more sophisticated ESD control in factory and assembly plants. The ANSI/ESD S20.20 [29] and other similar standards are being practiced quite widely in factories, and this has led to both significantly improved yields and reductions in ESD-related failures. However, non-ESD EOS stresses have a much wider variety of sources. Most times, an EOS failure shows much more extensive damage to the die than does ESD. An example of each is shown in Figure 3.8.

Figure 3.8a shows a more localized filament failure signature that is often associated with an ESD failure in MOSFETs. Figure 3.8b shows much more extensive damage resulting from a powered-on latched condition in a power management part. This is typical of the types of damage seen from EOS conditions and stresses. EOS damage results from transient stresses due to overvoltage in test or other electrical mishaps where the part is stressed beyond its absolute maximum temperature or

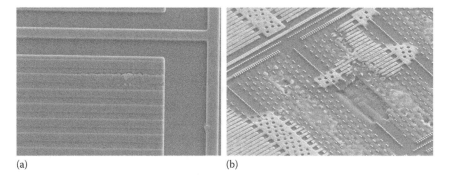

(a)   (b)

FIGURE 3.8   (a) Showing an ESD failure and (b) showing a non-ESD EOS failure.

power-handling capability [30]. While products are designed with ESD protection, mitigation of EOS most generally involves removing the high voltage/current that is serving as a source of the overstress from the environment where sensitive parts are being handled.

3.5   THERMAL AND DIFFUSION CONSIDERATIONS

Diffusion is a general phenomenon occurring in any state of matter and is a strong function of both temperature and time. The general diffusion relationship can be expressed as [31]:

$$\frac{\partial T}{\partial t} = \alpha \nabla^2 T \tag{3.7}$$

where:
   α is a constant
   ∇ is the Laplace operator

In the context of ESD/EOS occurring in microelectronics, thermal effects and diffusion leading to ESD/EOS damage will apply to semiconductor junctions, interfaces, surfaces, silicides, metals, and dielectrics used in the fabrication of chips. Because diffusion processes are thermally driven, the onset and ultimate failure by ESD/EOS is precipitated by thermal events, which causes diffusion leading to various failure modes. When a pulsed power event occurs, semiconductor junction breakdown can be initiated, causing diffusion-related junction degradation and ultimate failure. In silicides and metals, pulsed power events lead to various failure modes associated with diffusion, which increase the conductor resistance. Device gate oxide and dielectrics used around device wiring may suffer failure by dielectric breakdown. Weaknesses due to imperfect manufacturing can be the starting points for failure: for example, a semiconductor junction that is not truly homogeneous may lead to preferential overheating in a local area of the junction defect. Gate oxides may have point defects, imperfect interfaces, ionic contamination, or local thin areas that would tend to precipitate failure. Intra-level dielectrics can contain impurities, anomalies in their amorphous structure, and stresses and strains, which can all be weak areas susceptible to thermal and diffusion-related failures. In silicides and metals, dislocations, grain microstructure, surface defects, and current crowding can all be weak areas where failure modes precipitate. Pulsed power events and resulting diffusion can impact the materials in many ways. In devices, metallurgical junctions may not behave electrically as intended when diffusion occurs. For metals, diffusion can cause mechanical stresses on the metals and surrounding dielectrics, and metal void and hillock formation leading to resistance increases further aggravating the onset of failure. As described in Section 3.3 as well as the relationship in Equation 3.7, the pulse width influences the area of thermal rise. The thermal diffusion equation also applies to metal contacts and interconnect metallization. Very short ESD pulses may heat a small area, and thermal equilibrium and kinetic restraints will limit the volume in which diffusion can occur, so these events may be survivable. In longer-duration EOS events, the timescale may be long enough that kinetic barriers to diffusion are overcome and diffusion-related degradation and ultimate failure occurs. Because of the relationship between pulse width and thermal diffusion, large thermal gradients and related failure modes may occur even in narrow pulse ESD events. Even if a metallization system's performance is unaffected (no detectable parametric shifts) in ESD, a degradation in lifetime may occur because of subtle microstructural changes caused by diffusion. This is known as latent damage and can shorten device field lifetime. This will be discussed in more detail in subsequent sections.

3.6 METALLIZATION RELIABILITY AND ESD/EOS

When a metallization system is subjected to a pulsed stress event, it may fail resulting in an open circuit. Diffusion and the driving forces behind it are fundamental to understanding the failure modes. Fick's law of diffusion can be written as [32]:

$$J = -D\frac{\partial C}{\partial x} + \frac{DC}{kT}\left(Z^*eE - \frac{Q^*}{T}\frac{dT}{dx} - \frac{dU}{dx}\right) \tag{3.8}$$

where:
  J is flux of metal atoms
  D is diffusivity
  k is the Boltzmann's constant
  T is temperature
  Z^*e is effective charge
  E is the electric field
  Q^* is the heat of transport
  U is stress

The terms of the equation represent diffusion being defined by contributions from concentration gradients, electromigration (EM), thermo-migration, and stress gradients.

Although the laws of diffusion apply to all materials, in this section we focus on metals and their thermal and diffusion-related failure modes. In EM, sometimes called the *electron wind*, the momentum transfer from electrons to atoms results in atomic transport in the direction of electron flow, with voiding near the cathode and hillocks near the anode. The temperature gradients in thermo-migration favor the migration of atoms from cooler to hotter regions. In general, these mechanisms act together and may oppose or reinforce one another. For example, mass transport in EM always follows the direction of electron flow; however, concentration, thermal, and stress gradients may promote mass transport in the opposite direction. The direction in which metal atoms diffuse can provide insight into which failure mechanism is dominant. Regardless of which mechanisms dominate, diffusion can result in a void and ultimately an open circuit.

As diffusion is a thermally activated process, temperature is of particular interest. The activation energy for diffusion scales with the melting temperature (T_m) of the metal. Therefore, the higher the melting temperature, the more robust the metal will against diffusion in pulsed stress events.

3.6.1 DIFFUSION PATHWAYS

For a given polycrystalline metal, there are three primary diffusion paths: lattice (bulk), grain boundary, and surface diffusion. The activation energy for each is such that surface < grain boundary < lattice, and as a general rule of thumb, diffusion occurs at approximately 0.25, 0.50, and 0.75 of T_m for each case. Material choices and process integration can affect the overall metallization robustness by influencing both the T_m and diffusion pathways.

3.6.2 METALLIZATION RELIABILITY

To understand the effect of ESD and EOS pulses on metals in integrated circuits, several of the primary relationships pertaining to metallization reliability need discussion. By expanding on these relationships, the time dependence of these failure mechanisms becomes clearer.

In 1967, J.R. Black observed the relationship of mean time to failure (MTTF) due to EM, as a function of current density and temperature in a metal line [33]. This relationship is known as Black's equation. Black's equation for EM lifetime is:

$$\text{MTTF} = Aj^n e^{(Q/kT)} \tag{3.9}$$

where:
  A is a constant
  j is the current density
  n is a model parameter
  Q is the activation energy
  k is the Boltzmann's constant
  T is the temperature

The time to failure (typically defined as resistance increase of 20% in the metal) is proportional to microstructural properties:

$$\text{MTF} \propto \frac{S}{\sigma^2} \log\left(\frac{I_{111}}{I_{200}}\right)^3 \tag{3.10}$$

where:
    S is the grain size
    σ is the grain size distribution
    I are the X-ray intensity peaks for (100) and (200) planes in Bragg diffraction

As can be observed in the relationship described in Equation 3.10, larger grains with a tight size distribution and with surfaces oriented in the (111) plane are best. Metallization engineered in this way, along with higher melting temperatures, will lead to the highest reliability.

The driving force for stress migration (SM) is stress gradients that cause diffusion. Hooke's law describes the relationship between the concentration of lattice vacancies (proportional to diffusion) and the three-dimensional stress tensor:

$$\frac{\nabla C_L}{C_L} = -\frac{1}{B}\nabla\sigma \tag{3.11}$$

where:
    B, the bulk modulus, is a constant

The higher the bulk modulus of the material, the more resistant it is to stress gradient-driven diffusion. For thermo-migration, $\nabla T/T$ provides the driving force for diffusion, from cooler to hotter areas.

Each of these diffusion phenomena are contributors to metal failure under pulsed stress. Many metal failures occur in $1/t$ and $1/\sqrt{t}$ regimes of the Wunsch and Bell curve in Figure 3.4. In semiconductor junction breakdown resulting in a short, the current flow in the metal increases dramatically, and Joule heating and associated diffusion mechanisms are accelerated. With short power pulses, T can exceed melting temperature, but diffusion kinetics can be limited for narrower width pulses. Thermal energy can reach temperatures that can even melt tungsten, which has one of the higher melting temperatures for metals [34].

3.6.3 MATERIAL PROPERTIES AND PROCESS INTEGRATION OF AL AND CU

Some common metals are used in lines and vias, as shown in Table 3.1 [35], along with relevant material properties. For interconnects, Cu and Al are most commonly used, while W contacts and vias are common. For interconnects, Cu has more desirable material properties than Al, with higher melting temperature and mechanical strength and lower resistivity.

In terms of process integration, Cu lines and vias are formed by the damascene process. The damascene process is used because Cu cannot be dry etched by using

TABLE 3.1
Material Properties for Common Interconnect and Contact/Vias

Metal	T_m (°C)	Self-Diffusivity (cm²/s)	Young's Modulus (1E–11 dyn/cm²)	Resistivity (μΩ cm)
Al	660	1.71	7.06	2.65
Cu	1084	0.78	12.98	1.67
W	3400	0.04	41.1	5.65

Source: Murarka, S.P., *Metallization: Theory and Practice for VLSI and ULSI*, Butterworth-Heinemann, Oxford, 1992.

reactive ion etching (RIE). Al is blanket deposited and then RIE patterned, followed by dielectric fill and chemical-mechanical polishing (CMP). Cu damascene reliability weakness is surface due to the CMP process. Weak areas for Al are sidewalls due to their formation by RIE where defects and imperfect interfaces are more likely to occur.

3.6.4 Design Considerations

To maximize metal line reliability, narrow lines may be helpful [36]. As shown in Figure 3.9, as the metal grain size approaches line width dimensions, the grains start to span the entire line width, eliminating an important grain boundary diffusion path,

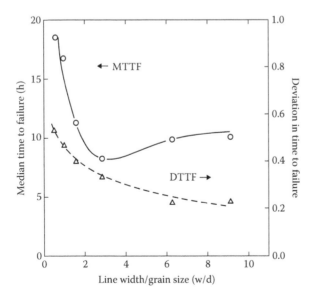

FIGURE 3.9 MTTF due to EM and deviation for Al interconnects. (Data from Cho, J. and Thompson, C.V., *Appl. Phys. Lett.*, 54, 2577–2579, 1989.)

and forcing bulk diffusion. Short lines of Blech length [37] balance EM and SM so there is no impact on wear-out lifetime. Interconnects will be a limiting factor in future ESD design, because of shrinking dimensions in line width pitch and height. Contacts and vias will also scale accordingly. Low-k BEOL dielectrics are also challenging because of their lower mechanical strength and different thermal properties. Metal line layouts susceptible to current crowding that can exacerbate i^{2}R heating should be avoided where possible [38].

3.6.5 ESD DAMAGE TO METAL STRUCTURES

During an ESD event, damage can occur to metal structures because of two general heating mechanisms. First, the metal itself can heat up because of current crowding and i^{2}R heating. This is often an indication that the metal line is not wide or robust enough to conduct the ESD energy safely. Because of the short duration of the normal ESD pulse, wide metal lines are not normally susceptible to this type of failure. In the second type of failure, heat is generated by nearby elements and then causes the metal structure to melt. This can easily occur because two of the most common interconnect metals, copper and aluminum, have lower melting temperatures than silicon or silicon dioxide (see Table 3.1). The melting point of silicon is 1414°C [39], thus many of the melt structures can be affected by heat diffusion even before the silicon device completely fails. This is particularly true for longer pulses such as those in an EOS event.

In a lower power pulse, the metallization may experience microstructural changes, such as recrystallization because of local melting and cooling, producing a film with more defects, larger grain boundary area, and different crystallographic orientations. The film has more Gibb's free energy than before (the sum of surface, interface, and strain energies). This latent damage does not significantly change the line performance; however, the wear-out field lifetime is significantly reduced because of the effects of EM and SM, which can lead to early field failure. In a higher pulse energy EOS event, the metal suffers irreversible changes in functionality due to melting, alloying reactions, void formation, and open failures from the Joule heating associated with the EOS event. Metallization failure is often associated with the device junction breakdown event, where permanent damage results in a short allowing increased current to flow into the metallization stack.

REFERENCES

1. L.R. Avery, "ESD Protection Structure Issues and Design for Custom Integrated Circuits," in *IEEE Custom Integrated Circuits Conference*, 1988.
2. D.C. Wunsch and R.R. Bell, "Determination of Threshold Failure Levels of Semiconductor Diodes and Transistors Due to Pulse Voltages," *IEEE Transactions on Nuclear Science*, vol. 15, no. 6, pp. 244–259, 1968.
3. N. Khurana, T. Maloney, and W. Yeh, "ESD on CHMOS Devices—Equivalent Circuits, Physical Models and Failure Mechanisms," in *Proceedings of the 23rd Annual Reliability Physics Symposium*, 1985.
4. P.S. Lim and W.K. Chim, "Latent Damage Investigation on Lateral Non-Uniform Charge Generation and Stress-Induces Leakage Current in Silicon Dioxides Subjected to Low-Level Electrostatic Discharge Impulse Stressing," in *Proceedings of the 7th International Symposium on the Physical and Failure Analysis of Integrated Circuits*, Singapore, 1999.

5. I.-C. Chen, J.Y. Choi, and T.-Y. Chan, "The Effect of Channel Hot-Carrier Stresing on GAte-Oxide Integrity in MOSFET's," *IEEE Transactions on Electron Devices*, vol. 35, no. 12, pp. 2253–2258, 1988.
6. J. Wu, P. Juliano, and E. Rosenbaum, "Breakdown and Latent Damage of Ultra-Thin Gate Oxides Under ESD Stress Conditions," in *EOS/ESD Symposium*, 2000.
7. A. Ille, W. Stadler, A. Kerber, T. Pompl, T. Brodbeck, K. Esmark, and A. Bravaix, "Ultra-Thin Gate Oxide Reliability in the ESD Time Domain," in *EOS/ESD Symposium*, 2006.
8. S. Malobabic, D.F. Ellis, J.A. Salcedo, P. Zhou, J.-J. Hajjar, and J.J. Liou, "Gate Oxide Evaluation under Very Fast Transmission Line Pulse (VFTLP) CDM-Type Stress," in *Proceedings of the 7th International Caribbean Conference on Devices, Circuits, and Systems*, Mexico City, 2008.
9. A. Ille, W. Stadler, T. Pompl, H. Gossner, T. Brodbeck, K. Esmark, P. Riess, D. Alvarez, K. Vhatty, R. Gauthier, and A. Braviax, "Reliability Aspects of Gate Oxide under ESD Pulse Stress," in *EOS/ESD Symposium*, 2007.
10. R. Duschl and R.-P. Vollertsen, "Voltage Acceleration of Oxide Breakdown in the dub-10 nm Fowlewr-Nordheim and Direct Tunneling Regime," in *IEEE International Reliability Workshop*, 2005.
11. J. Reiner, "Latent Gate Oxide Defects Caused by CDM-ESD," in *EOS/ESD Symposium*, 1995.
12. C. Ito and W. Loh, "A New Mechanism for Core Device Failure during CDM ESD Events," in *EOS/ESD Symposium*, 2006.
13. V.M. Dwyer, A.J. Franklin, and D.S. Campbell, "Electrostatic Discharge Thermal Failure in Semiconductor Devices," *IEEE Transactions on Electron Devices*, vol. 37, no. 11, pp. 2381–2387, 1990.
14. J.C. Reiner, T. Keller, H. Jaggi, and S. Mira, "Impact of ESD-Induced Soft Drain Junction Damage on CMOS Product Lifetime," in *Proceedings of the 8th IPFA*, Singapore, 2001.
15. G.A.M. Hurkx and N. Koper, "A Physics-Based Model for the Avalance Ruggedness of Power Diodes," in *Proceedings of the 11th International Symposium on Power Semiconductor Devices*, 1999.
16. D.M. Tasca, "Pulse Power Failure Modes in Semiconductors," *IEEE Transactions on Nuclear Science*, vols. NS-17, no. 6, pp. 364–372, 1970.
17. M. Lutzky, E.B. Dean, and M.C. Petree, "Modeling Second Breakdown in PN Junctions with NET-2," *IEEE Transactions on Nuclear Science*, vols. NS-22, no. 6, pp. 2645–2649, 1975.
18. D.L. Blackburn and D.W. Berning, "An Experimental Study of Reverse-Bias Second Breakdown," in *International Electron Devices Meeting*, 1980.
19. G. Krieger, "Thermal Response of Integrated Circuit Input Devices to an Electrostatic Energy Pulse," *IEEE Transactions on Electron Devices*, vols. ED-34, no. 4, pp. 877–882, 1987.
20. S.H. Voldman, *ESD Failure Mechanisms and Models*, West Sussex: John Wiley & Sons, 2009.
21. S. Voldman, "Electrostatic Discharge (ESD) and Failure Analysis: Models, Methodologies and Mechanisms," in *Proceedings of the 9th IPFA*, Singapore, 2002.
22. A. Amerasekera, L. van Roozendaal, J. Bruines, and F. Kuper, "Characterization and Modeling of Second Breakdown in NMOST's for the Extraction of ESD-Related Process and Design Parameters," *IEEE Transactions on Electron Devices,* vol. 38, no. 9, pp. 2161–2168, 1991.
23. C. Diaz, C. Duvvury, and S.-M. Kang, "Thermal Failure Simulation for Electrical Overstress in Semiconductor Devices," in *IEEE International Symposium on Circuit Systems*, 1993.
24. D.A. Neamen, *Semiconductor Physics & Devices*, Second Edition, Boston, MA: Irwin/McGraw-Hill, 1997.

25. F. Farzan, A. Appaswamy, A.A. Salman, and G. Boselli, "Overshoot-Induced Failures in Forward-Biased Diodes: A New Challenge to High-Speed ESD Design," in *IEEE International Reliability Physics Symposium*, 2013.

26. M. Lee, "ESD Induced Damage and Hot-Carrier Reliability of NMOS and PMOS Transistors," in *Proceedings of the 11th Biennial University/Government/Industry Microelectronics Symposium*, Austin, TX, 1995.

27. R.L. Davies and F.E. Gentry, "Control of Electric Field at the Surface of P-N Junctions," *IEEE Transactions on Electron Devices*, vol. 11, no. 7, pp. 313–323, 1964.

28. S. Dabral and T.J. Maloney, *Basic ESD and I/O Design*, New York: John Wiley & Sons, 1998.

29. Electrostatic Discharge Association (ESDA), *ANSI/ESD S20.20 Protection of Electrical and Electronic Parts, Assemblies and Equipment*, Rome, Italy: The ESD Association, 2014.

30. C. Diaz, S.M. Kang, and C. Duvvury, "Electrical Overstress and Electrostatic Discharge," *IEEE Transactions on Reliability*, vol. 44, no. 1, pp. 2–5, 1995.

31. P. Dawkins, *Paul's Onlline Math Notes*, 2003. [Online]. Available: http://tutorial.math.lamar.edu/Classes/DE/TheHeatEquation.aspx. [Accessed February 9, 2015.]

32. M.E. Glickman, *Diffusion in Solids: Field Theory, Solid-State Principles, and Applications*, New York: John Wiley & Sons, 2000.

33. J.R. Black, "Electromigration—A Brief Survey and Some Recent Results," in *Proceedings IEEE International of the Reliability Physics Symposium*, Wahsington, DC, 1968.

34. A.J. Walker, K.Y. Le, J. Shearer, and M. Mahajani, "Analysis of Tungsten and Titanium Migration during ESD Contact Burnout," *IEEE Transactions on Electron Devices*, vol. 50, no. 7, pp. 1617–1622, 2003.

35. S.P. Murarka, *Metallization: Theory and Practice for VLSI and ULSI*, Butterworth-Heinemann, Oxford, 1992.

36. J. Cho and C.V. Thompson, "Grain Size Dependence of Electromigration Induced Failures in Narrow Interconnects," *Applied Physics Letters*, vol. 54, no. 25, pp. 2577–2579, 1989.

37. I.A. Blech, "Electromigration in Thin Aluminum Films on Titanium Nitride," *Journal of Applied Physics*, vol. 47, no. 4, pp. 1203–1208, 1976.

38. J. Lienig, "Interconnect and Current Density Stress: An Introduction to Electromigration-Aware Design," in *Proceedings of the 7th International Workshop on System-Level Interconnect Predictions*, San Francisco, CA, 2005.

39. Royal Society of Chemistry, *Periodic Table*, Royal Society of Chemistry, December 2014. [Online]. Available: http://www.rsc.org/periodic-table/element/14/silicon. [Accessed January 30, 2015.]

4 ESD, EOS, and Latch-Up Test Methods and Associated Reliability Concerns

Alan W. Righter

CONTENTS

4.1 INTRODUCTION

This chapter will cover the primary test methods used to evaluate integrated circuit (IC) robustness to electrostatic discharge (ESD), transient overstress, and latch-up events. Additionally, reliability issues relating to these short-duration electrical events will be described.

There are three main mechanisms producing high-energy/short-duration electrically induced damage in ICs: ESD (generally less than 1 µs in duration), the more general electrical transients (those of longer duration from microseconds to milliseconds), and latch-up.

ESD represents a short-duration, high-current discharge, resulting from either a triboelectric (two dissimilar materials coming together and separating) effect or an inducing electric field. For ICs, the discharge can take place through either a circuit path between pins of an IC or a sudden discharge from a single pin of a charged IC. Typically, these different ESD discharges, described by two main component-level "qualification" ESD test models, are under 1 μs in total duration. These ESD models (human body model [HBM] and charged device model [CDM]) assume that the IC is unpowered when such an event occurs. It is important that to ensure reliability and protection from these events, alternate current paths are designed into the IC or system to direct the high values of ESD current (and high path voltage) away from critical internal physical areas of the circuit such as gate oxide of transistors, small input P–N junctions, and narrow metal lines, especially those lower-level metal lines, which may be affected by topography of polysilicon for non-planarized processes. ESD events and models are described in Sections 4.2.1 through 4.2.3.

Other electrical transients (separate from ESD) refer to those caused by any non-triboelectric or non-electric field-induced electrical overstress (EOS) mechanism, and particularly those over 1 μs in duration. Although EOS damage can result from a direct electrical transient applied to one or more input/output (I/O) pins of a device, it can also result from supply transients due to unintended connection situations, either internal to or external to the device, which sensitize the device to localized energy that can result in overstress [1,2]. Resulting EOS-like damage in electronic circuitry often results in severe (often observable optically) physical damage, such as melted metal, fusing and/or discoloration of silicon junctions/metal, and even blown holes in silicon. EOS-like damage can often leave a trail indicating the path taken through the IC, and the electromigration signature often provides a clue to its direction of stress. There is no single test for component-level overstress, but system-level ESD tests and system-level stresses such as IEC 61000-4-4 (Electrical Fast Transient/Burst) [3] and 61000-4-5 (Surge Immunity) [4] can model some types of stress resulting in severe EOS damage in devices.

Latch-up occurs in CMOS ICs from overvoltage resulting from an applied overcurrent (to I/O pins) or overvoltage (to supply pins), either of which creates an unintended interaction and subsequent current amplification involving parasitic bipolar structures (inherent in the CMOS architecture) that result in a regenerative, positive amplification current path from a supply to a ground. Uncontrolled latch-up can result in EOS damage. Latch-up, its reliability effects, and prevention will be discussed further in Section 4.2.4.

Section 4.3 will discuss reliability of ESD tested product, including the snapback mechanism. Finally, ESD and EOS impacts on reliability, along with suggested guidelines for minimization, will be discussed in Section 4.4.

4.2   DEVICE-LEVEL ESD TESTS/STANDARDS

There are three widely used ESD models describing different energy transmissions of the ESD event for IC components: the HBM, the CDM, and the machine model (MM).

4.2.1   HUMAN BODY MODEL

The HBM of ESD described in ESDA/JEDEC JS-001 [5] electrically models the transient stress resulting from a person making physical contact with a pin of the IC with another pin or pin group of the IC held at a lower potential or ground. The basic electrical human body circuit of a person described in this model consists of 100 pF shunt capacitor to ground, a 1500 Ω resistor in series with the device under test (DUT) path, and a small inductance, generally 5–10 nH, in series with the resistor. The HBM applies a charge voltage (in hundreds or thousands of volts) discharged from the shunt capacitor through the 1500 Ω/DUT path. The peak currents generated through the device path are the applied voltage divided approximately by the $(1500 + R_{DUT})$ ohm resistance. The resulting HBM waveform is a "double exponential" shape with defined rise and fall times. For example, a 2000 V HBM applied voltage has a peak current of approximately 1.33 A, and this peak current is reached anywhere from 2 to 10 ns after the voltage begins to be applied. In the model, after the peak current is reached, this current decreases until it reaches 37% of its value within 150 ns after the voltage is applied. The HBM circuit description and waveform are shown in Figures 4.1 and 4.2.

During a HBM ESD test, different pin stress combinations are applied, which represent the discharge paths through the device pins. There are two variations of pulse combinations represented in JS-001. One variation is the "Table 2B" list of pin combinations, which consists of the following:

1. I/O pin to a supply (a supply can be either a circuit power or ground), all other pins floating: One positive and one negative zap is applied to an I/O pin of the IC with respect to a pin/pin group of a power supply or ground rail that is connected to tester ground. During each test combination in this set, all other I/O pins and supplies are floating.
2. Power rail to power rail, all other pins floating: One positive and one negative zap is applied to one power supply or ground potential with respect to another power supply or ground held at ground. Every such power/ ground domain combination of the device must be tested. During each test

Human body model (HBM)

- Simulates the discharge from the finger of a standing person
- RLC = 1.5 kΩ, 7.5 μH, 100 pF ⟹ An ideal current source

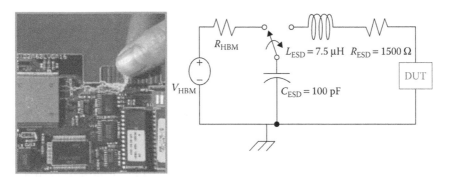

FIGURE 4.1   Human body model (HBM) circuit description. (a) Simulates the discharge from the finger of a standing person. (b) For an ideal current source RLC = 1.5 kΩ, 7.5 μH, 100 pF.

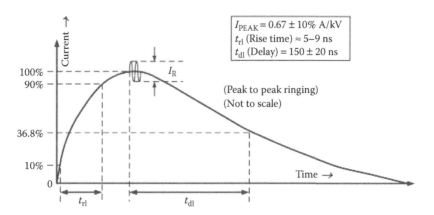

FIGURE 4.2 HBM ESD pulse waveform into a short—circuit: Typical test levels-500 V to 4000 V.

combination in this set, all I/O pins and supplies not in the particular combination are floating.

3. I/O pin to group of I/O pins, all other supplies/grounds floating: Each I/O pin is zapped once positively and once negatively with respect to a ganged connection of the remaining non-supply pins. During each test combination in this set, all supplies are floating.

A second (more concise) variation of pulse pin combinations in JS-001 is listed in "Table 2A." This list of pin combinations is designed to minimize the number of "cumulative" or repetitive pin zap combinations while still preserving sufficient pin combination coverage. Specifically, use of the Table 2A combinations allows the following:

- I/O pins only need to be stressed to the supplies and grounds that provide the power to the I/O pin circuit. This means I/O pins do not need to be stressed to supplies not providing power or ground to the I/O circuit, and those supplies can be left floating.
- A special group of I/O pins called *coupled pin pairs* are those I/O pins that have a circuit connection to each other. Examples are HDMI, LVDS, and amplifier inputs. For these coupled pin pairs, these pins are stressed only to each other and not to the group of remaining I/O pins, which can be left floating.
- For supply-to-supply stressing, only positive stressing needs to be done. As every supply is stressed positively to every other supply, this guarantees that every supply would be stressed positively and negatively to every other supply.
- It is allowed to replace a negative zap on an I/O pin with respect to a grounded supply with a positive zap on the supply with respect to the

grounded I/O pin. It has been shown that HBM tester parasitics can negatively affect the path to the grounded supply, and reversing the direction of the zap provides the same stimulus but with reduced effects from the parasitics.

For HBM qualification testing, every zap combination must be applied to at least three parts per voltage step, to all circuit pin combinations over a range of test voltages. Testing can start at any voltage, but in practice the voltage range is generally in 500 V steps from 500 to 4000 V, or more. Lower-voltage levels are also applied, and these lower levels can be evaluated in finer voltage steps. The JS-001 standard specifies classification levels from 125 to 8000 V. Each pin combination to be zapped must be zapped once positively and once negatively with respect to another pin/supply tied to ground, with all other pins/supplies floating.

Prior to and following the HBM stressing, the ICs are tested to the same full datasheet electrical test, to detect post-zap changes in leakage current and/or IC functional failure.

IC HBM damage is primarily thermal in nature. Chapter 3 gives examples of HBM failures. For example, the high current in reverse-biased avalanching junctions causes localized heat buildup. As the temperature increases, the doping concentration difference across the junction decreases. Eventually, the junction temperature gets so high that the doping concentration difference becomes very small, and the melting point of silicon is reached. Often metal contacts make contact with the silicon surface near the junction, and the melting point of metal is lower than the silicon. The metal expands and spikes through the junction causing damage. Figure 4.3 shows an example of HBM damage from drain to source through a gate finger of an MOS I/O transistor.

FIGURE 4.3   Example of contact spiking/junction damage from a HBM ESD event.

4.2.2   Charged Device Model

The CDM of ESD is the real-world ESD model representing ESD events from automated manufacturing and handling of IC components [6]. CDM was first discovered in the 1980s by AT&T and represents the discharge from and out of an IC (which has acquired a positive or negative potential) into a conductor, which is a lower potential or ground. It has been estimated that over 95% of ESD events result from a CDM event.

One example of CDM charge/discharge in manufacturing occurs when an IC sliding along a conveyor of a poorly grounded test handler triboelectrically develops a positive or negative potential during the sliding along the conveyor travel path. A pin of this charged IC then comes into contact with a well-grounded portion of the tester handler, and a discharge from the IC substrate/package through the contacted pin to ground results.

The energy-time envelope of this CDM event, shown by representative current waveforms, shows that it occurs much faster than the HBM or MM event and has a peak current that is significantly higher than that of either HBM or MM. Figure 4.4 shows the comparison of CDM, HBM, and MM waveform time profiles and example peak currents.

It is important to note that the source of the CDM discharge energy is the device itself, in contrast to the HBM event, which comes from an external human source. The IC capacitance, largely due to the package area and mold compound thickness, is the main source of energy. The CDM discharge envelope has positive and negative excursions from the zero charge value, similar to a damped sinusoidal wave. The first peak current is the worst-case portion of the discharge event, and this first peak occurs anywhere from 100 to 700 ps from the start of the discharge, depending on the external pin ground RLC (resistance/inductance/capacitance) connection, and internal package RLC. The discharge event typically decays to near zero amps in 1–3 ns, again depending on the capacitance of the IC.

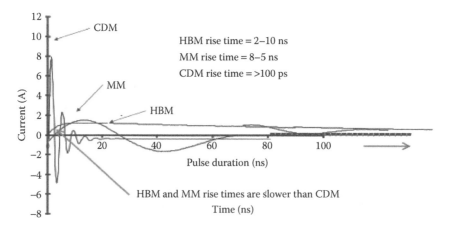

FIGURE 4.4   Comparison of HBM, CDM, and MM stressing waveform timescales/peak currents.

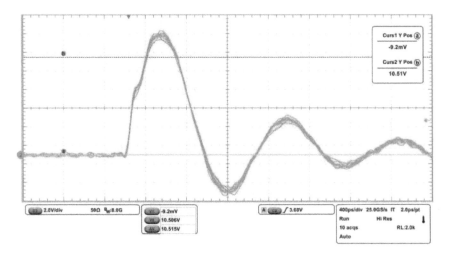

FIGURE 4.5 Example of +500 V CDM ESD discharge waveforms taken with a high bandwidth digital oscilloscope.

Figure 4.5 shows a set of waveforms (with a 6 GHz oscilloscope) taken using a JEDEC-style large coin module with the field plate set to a 500 V field-induced charge voltage.

The electrical model for the field-induced CDM ESD event is represented by a combination of tester and die capacitances to the field plate and tester chassis ground [7].

A CDM test is best illustrated with a CDM tester schematic representing a CDM tester with a representative device and its various capacitances, as shown in Figure 4.6. The device is placed with its package such that its pins face upward and that its package makes flat contact with a dielectric sheet covering the "field plate," or plate that connects to the high-voltage supply used to induce the voltage to the device. The device is then charged through the field plate to a set voltage and then a discharge "probe assembly" probe is brought near to a pin to discharge it. Every pin of the device receives positive and negative charge/discharge cycles. A minimum of three devices are tested at a particular voltage level.

In the JEDEC Charged Device Model Standard JESD22-C101, a minimum of three ICs must be subjected to ESD discharge at a particular voltage, with the voltage steps for each group of three devices being 125, 250, 500, and 1000 V. Prior to and following the zapping, the ICs are subjected to a full datasheet electrical test, to detect parametric changes or functional failure.

Oxide damage in transistor gate oxides (particularly small input gates and internal cross-domain circuit nodes between two different domains) and small non-supply pin capacitors connected with minimum resistance to I/O pins are common modes of failure in CDM ESD, as well as junction damage in small area junctions. Chapter 3 gives examples of these failures. Figure 4.7 shows a CDM oxide damage failure, showing a "mouse bite" effect at the edge of the polysilicon gate with respect to the

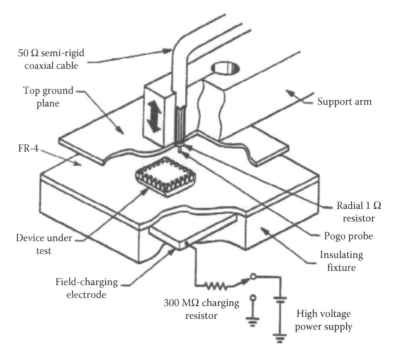

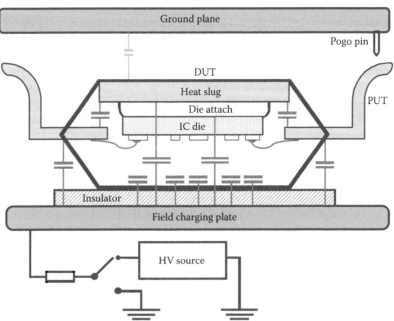

FIGURE 4.6 CDM tester schematic and device capacitances. (Data from Chaine, M. and Etherton, M., *Predictive Circuit Simulation of Charged Device Model ESD Events, Tutorial at 2012 ESD Symposium*, 2012.)

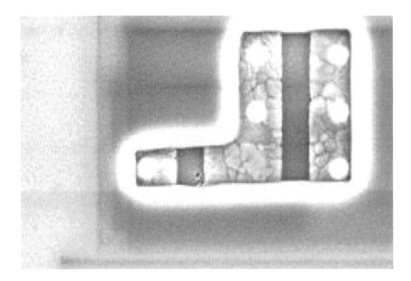

FIGURE 4.7   Oxide damage resulting from a CDM ESD event on an input NMOS transistor gate.

source of an NMOS transistor [8]. Another CDM failure mode is metal rupture of thin lower-level metal lines running over minimum-spaced polysilicon lines.

As of press time, a new CDM test standard co-authored by both the EOS/ESD Association and JEDEC, ANSI/ESDA/JEDEC JS-002-2014, has just been published [8]. JS-002 replaces the existing ESDA and JEDEC CDM standard documents and successfully harmonizes the ESDA and JEDEC CDM test platforms into a single test platform. This new platform uses the JEDEC-style small (6.8 pF) and large (55 pF) verification modules and the JEDEC-style FR4 dielectric atop the field plate but removes ferrite components from the discharge probe (ground plane) test head, which had resulted in inaccuracies in the CDM waveform for higher frequencies. This new standard has five "test condition" classification voltages that are set on the tester by calibrating the tester voltage to achieve defined peak current level ranges for each of the five separate test condition, representative of the JEDEC peak currents realized from JEDEC tester set voltages of 125, 250, 500, 750, and 1000 V, respectively. This new joint standard also requires initial tester calibration and annual servicing/re-verification to use a high bandwidth (≥6 GHz) oscilloscope (1 GHz oscilloscopes are still allowed for periodic waveform checking).

4.2.3   MACHINE MODEL

Recently demoted to be a characterization-only ESD test and not used for component ESD qualification, the MM of ESD [9] was developed in Japan and was intended to electrically model the waveform created when a person holding a metal object touches a pin of a device. For MM, the model circuit capacitance to ground is larger than for HBM (200 pF compared to 100 pf for HBM), and there is very little to no

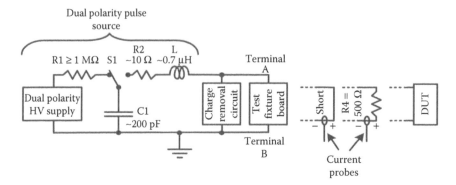

FIGURE 4.8   Example of MM ESD circuit. (Data from EOS/ESD Association, *ANSI/ESD/ STM 5.2 Machine Model, Component Level*, 2012.)

series resistance, compared to HBM. This model was once believed to be a worst-case form of HBM; however, observed waveforms in actual manufacturing discharges are damped oscillations but with much faster waveform period, similar to the CDM waveform. An example of MM circuit and waveform is shown in Figure 4.8.

The MM zapping procedure follows the same pin stress combinations as the HBM, but the applied voltages are lower, typically 25, 100, 200, and 400 V for each group of three devices, primarily because of the low MM series resistance (see Figure 4.9).

Damage from MM ESD events correlates quite well to damage resulting from HBM events. However, because of the variation of the low actual resistances of the circuit model implementation in test equipment, the MM tester parasitic effective capacitance and inductance gives so much variation to the waveform that no two testers produce the same results (the same failure locations but different failure voltages). Also, it has been shown that the type of discharge (a charged conductor discharging

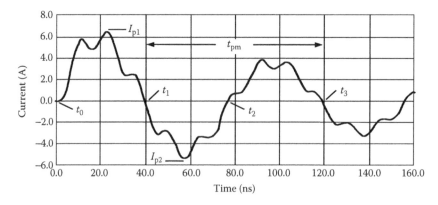

FIGURE 4.9   Example of 400 V MM ESD waveform. (Data from EOS/ESD Association, *ANSI/ESD/STM 5.2 Machine Model, Component Level*, 2012.)

into the part) in actual ungrounded machine measurements produces a waveform very similar to CDM, so this model is not representative of ungrounded "machines" [10]. For these reasons, the MM has been decommissioned to a characterization-only test method and is not to be used for qualification testing of product according to most international standards bodies (ESDA, JEDEC, AEC, JEITA). The HBM and CDM ESD test methods are the two product qualification component-level ESD test methods.

A variation of the MM test has been used to apply the MM pulse to a powered device. In this test, I/O and power pins are zapped in turn with positive and negative pulses. The observed damage can result from a transient form of latch-up (described in the next section) or EOS-like damage, and not necessarily damage resulting from an ESD event.

4.2.4 LATCH-UP

Latch-up occurs in CMOS ICs from interactions between parasitic bipolar structures (inherent in the CMOS architecture) that create a regenerative, positive amplification current path [11]. It occurs as a result of beta multiplication (amplification) of current in a parasitic pnpn path created from diffusions and substrate/wells of NMOS and PMOS transistors. If not controlled, latch-up results in destruction of the affected transistors and can cause EOS-like damage to supply and ground metal to these structures. This results from insufficient isolation of these opposite polarity transistors in the IC layout, creating a layout-dependent parasitic conduction path. Activation of this conduction path results from an overvoltage or overcurrent condition on either the supply or I/O node and causes a high current to flow, which can cause permanent IC damage if the excess current is not shunted away or otherwise limited into the circuit.

An example of a CMOS inverter and its electrical structure is shown in Figure 4.10.

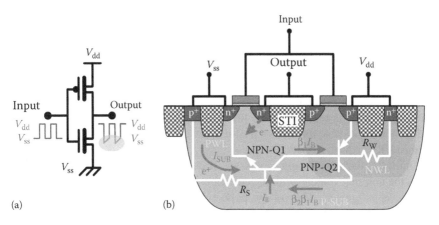

(a)   (b)

FIGURE 4.10 Example of a CMOS inverter (a) and cross section of its parasitic bipolar circuit (b) that creates the latch-up event.

The CMOS technology contains a pnpn sandwich of layers, and the figure shows a connection of a parasitic pnp and a parasitic npn transistor. The source of the n-channel transistor forms the emitter of a parasitic lateral npn transistor, the substrate forms the base of the transistor, and the n-well forms the collector. The source of the p-channel transistor forms the emitter of a parasitic pnp transistor, the n-well forms the base, and the p-type substrate forms the collector. Note the two resistors, R_{well} and R_{sub}. In order for latch-up to initiate, one of the normally zero-biased junctions in the structure must become forward biased. In order for this to occur, current must exist in one of the resistors between the emitter and base of one of the two bipolar structures. This current can come from a variety of causes such as an application of a high-spike voltage that is larger than the supply voltage to an I/O terminal, or improper sequencing of the power supplies. Testers must be properly programmed to apply input pin voltages no higher than the supply voltage, and multiple V_{dd} power supplies (for ICs with separate I/O) and core logic or large circuits requiring the use of multiple V_{dd} pins must be programmed to turn on in the right order.

Latch-up is more likely to occur in circuits as the substrate and well doping concentrations are made lighter (increasing the resistance), as the well is made thinner (also increasing the resistance), and as the IC transistor threshold voltages are scaled downward, resulting in less gate voltage required to turn on. The β (gain or ratio of collector to base current) of the two bipolar transistors also tends to increase as technologies scale downward, and the product of the betas need only to be above unity for amplification and possible latch-up to occur. Special protection structures such as diode-resistor networks on I/O pads can help reduce this effect as currents into or out of the chip are directed away from circuitry.

The JEDEC JESD78 latch-up test standard [12] outlines latch-up test procedures for all I/O pins and power supply pins with stressing current (I/O) and voltage (power supply) pulses, respectively.

For I/O pins, a current pulse test applies a sequence of increasing trigger current pulses, typically starting at 10 mA, and continuing until $I_{norm} + 100$ mA or $1.5 \times I_{norm}$ (whichever is greater; I_{norm} is defined as nominal operating current). The stress voltage during each current pulse is clamped at 1.5 times the maximum operating voltage (typically $V_{dd} + 10\%$). The trigger current pulse can last anywhere from 10 µs to 5 ms (which is the preferred time to give the maximum stress). After this stress pulse, the nominal voltage operating current is measured and compared with nominal pre-stress operating current. If this post-latch-up test operating current is more than 1.4 times the nominal current at a given stress level, that test level fails. If the clamping voltage is reached before the maximum current is applied to the device, and there is no leakage increase or post-test electrical failure, latch-up for that device is considered to have passed.

Similarly, the power supply pins are stressed with a voltage pulse that is 1.5 times the nominal voltage, with the current clamped at $I_{norm} + 100$ mA or $1.5 \times I_{norm}$ (whichever is greater). The voltage pulse typically lasts for up to 5 ms. After this stress pulse, the nominal voltage operating current is measured and compared with nominal pre-stress operating current. If this post-latch-up test operating current is more than 1.5 times the nominal current, the test fails.

4.2.4.1 Latch-Up Testing and EOS Concerns

There are cases for different processes, notably high-voltage (30 V or above) and very low–voltage (under 1.2 V) processes, where the maximum stress voltage (MSV) of 1.5 times maximum operating voltage represents an EOS condition to the part. This is because the breakdown voltage of the process junctions is less than this value. For example, for a typical 60 V BCDMOS process, if the (1.5 times $V_{dd(nom)}$ + 10%, or $V_{max(op)}$) voltage, which is 99 V, were to be applied to a cathode of a high-voltage junction where the anode voltage was held at zero volts, this voltage would be above the breakdown voltage of the junction. This would likely result in an EOS condition. In these cases, an MSV must be determined that represents the maximum safe voltage to be applied to such junctions, which still gives adequate stress to the part to detect latch-up-prone circuit architectures while not exhibiting the part to an EOS condition. Process characterization should be done to determine a safe maximum voltage that can be applied in the latch-up test so it does not represent an EOS condition for the part.

Also, a similar case occurs for CMOS processes with very low nominal voltage, below 1 V or so. The I/O devices in these processes, particularly those with very small I/O devices, cannot stand high currents and have high on-resistances that cause voltage drops across bipolar junctions to be more sensitive with greater potential for failure. Careful attention to guard ringing and separation, as well as placing NMOS in separate deep n-well (so the NMOS is in a separate p-tub within the deep n-well), largely solves the latch-up issue.

4.2.4.2 Transient Latch-Up

The traditional latch-up test described in the JEDEC JESD78 document has minimum trigger pulse rise times and durations of 5 and 10 μs, respectively. These times are generally much longer than component ESD test pulses, and recent work has shown that latch-up can also occur for shorter, faster rise time pulses. An ESD Association Technical Report, ESD TR5.4-04-13 [13], describes the nature of these transient latch-up stimulus events, describes how they can be simulated, and shows case studies of transient latch-up resulting from a number of different types of stress. Care must be taken to evaluate if very short high-voltage pulses can trigger latch-up in applications.

4.3 RELIABILITY OF ESD TESTED PRODUCT

In each of the aforementioned component-level ESD test methods, ESD testing is considered destructive to the tested part, so devices tested for ESD are not to be used as commercially viable for sale. However, if sample size is limited, devices used for each test that pass may be retested at higher voltages for the same ESD test until first failure from post-electrical testing occurs. However, it is important to understand the nature of the phenomenon that transistors have to withstand ESD events, described in the next section.

4.3.1 Snapback

In the 1980s, a phenomenon called *snapback* was observed in CMOS n-channel transistors. Snapback is a phenomenon unique to MOS circuitry [14].

A snapback example can be illustrated by a simple CMOS inverter. If a signal of logic 0 drives the inverter, the p-channel is turned on and the n-channel is turned off. Increasing V_{dd} and V_{in} causes the voltage at the n-drain to increase. Sufficiently high voltage causes avalanche breakdown, and current is injected into the p-substrate. If the supply and input voltages are held steady, the substrate current will increase until the source/substrate junction is forward biased. This junction will then inject carriers into the substrate in the direction of the drain due to field-enhanced injection. The increased current decreases the drain potential and avalanching of the drain/substrate junction largely disappears except for the region directly facing the source/substrate junction. The focused injection continues from the substrate/source junction. This effect lowers the effective resistance of the base region of the n-transistor corresponding to the channel region. The n-channel is not turned on in the normal sense, but it does conduct current similar in operation to an n-transistor with a threshold voltage of 0 V. The effect on test is that application of a voltage higher than the snapback-inducing voltage will result in the IC suddenly consuming much more current with the supply voltage scaling back to a lower snapback-sustaining voltage. This large I_{dd} results from the current increase in the n-transistor, which should normally be off. Logic function will not be interrupted if the resulting current does not overdrive the p-transistor. The snapback mechanism will terminate when the p-channel load turns off and the n-channel turns on.

A high incidence of HBM ESD failures occurs in NMOS transistors, due to the snapback mechanism, which is more prevalent in NMOS because of the increased efficiency of the parasitic bipolar npn device over the corresponding PMOS parasitic pnp. Such snapback is not as prevalent in PMOS transistors but is known to happen for smaller geometry processes. Addition of a normally off NMOS transistor that is ESD hardened and capable of quick operation into snapback (between I/O pad and active circuitry), together with series resistance placed between drains of the ESD and signal NMOS, helps steer damaging ESD currents away from NMOS signal driver logic and input gates to the power rails that are ESD protected.

4.4   IC PRODUCT RELIABILITY TESTS AND ESD/ EOS/LATCH-UP CONCERNS

IC reliability testing involves exposure of ICs to various "acceleration" factors of voltage, temperature, moisture, shock/vibration, and so on to check for the presence of defects that manifest and result in early life manufacturing failures. To set up and perform these tests, various handling and operation functions are done to prepare and condition the parts. It is these steps that can inadvertently result in device failure due to overstress.

One example is the High Temperature Operating Life (HTOL) test, typically performed on a statistically large sample of ICs. This would also include tests such as Early Life Failure (ELF) and Highly Accelerated Stress Test (HAST), which generally use the same boards. Voltage conditions for this test are generally 10%–20% above the nominal supply voltage, with elevated temperature from 125°C to 150°C, for anywhere from 48 to 1000 (or more) hours. The intent of these tests is to detect any ICs that have latent defects such that they fail during the time of the test. During a "normal" accelerated test run, the conditions are such that normally no devices would fail this test. However, various conditions can happen that cause overstress. For example, power supplies in burn-in environments are noisy in that transient excursions (spikes) from the intended supply voltage can randomly occur, which may be 1–2 V over the set voltage. The set voltage during the HTOL test should have enough headroom to not exceed the absolute maximum rating (AMR) voltage level of the product. However, such transients on the order of hundreds of nanoseconds, to tens of microseconds, can occur. At the high operating temperature, the designed turn-on voltages of on-chip ESD protection devices can be lowered and thus will turn on to shunt the current in ESD operating mode, but because of the long duration of these transients, the devices can be overstressed to failure. To protect against this, transient voltage suppressor (TVS) or Zener devices connected between test board supply and ground are placed close to the connection entry point of the supply into the board. If a supply voltage transient occurs, the TVS/Zener devices will turn on with a low resistance and shunt the current away. Additionally, Schottky diodes are also recommended to be placed, which turn on for voltage transients in the reverse direction. It is recommended also to add a high–power, low-value series resistor between this TVS/ Schottky ESD protection and the IC devices themselves, to ensure that the TVS or Schottky devices will turn on before the ICs.

A second reliability issue concerns the reliability test and electrical tester test boards used for the IC reliability tests. Care must be taken with these boards to prevent the board (and IC) from acquiring a voltage significantly raising or lowering its potential with respect to ground. In the charged board event (CBE) [15], the total board capacitance often is many times that of a single IC capacitance, due to the large PCB capacitance between supply and ground planes and, for reliability test boards, many such IC devices placed in parallel between supply and ground. A discharge of a charged board to a grounded node can result in very large currents that can focus through the IC power and ground routing and can cause failure of the IC. This board-acquired CBE voltage can cause damage at a lower voltage, compared to the CDM voltage level of just the component. Care must be taken to place insulators, such as plastic connectors, large capacitors, or inductors away from susceptible ICs, particularly those with many V_{dd}/V_{ss} pins, on larger boards. ICs with large numbers of power supply and ground pins can be especially vulnerable because of the very low resistance from supply and ground routing through the IC subjecting the IC metal to very large currents. Supply and ground PCB "power planes" should be spaced further apart if possible to reduce the overall PCB board capacitance. Adding TVS devices across supply/ground planes of the IC allows safe rapid discharge of voltage through each TVS preventing internal board circuit failure.

4.5 SUMMARY

IC failure from overvoltage/overcurrent can occur from three different methods of externally induced electrical stimuli, namely ESD, transient events that lead to EOS damage, and latch-up. Physical descriptions, test methods, and failure modes/damage of each of these methods have been described in this chapter. This has led to a discussion of how reliability can be potentially affected by these different mechanisms.

Care must be taken at the IC production/bench test, IC reliability test, system board assembly, and system connection steps to put protection and control methodologies in place to minimize the occurrence of events leading to these failure modes.

REFERENCES

1. A. Righter, E. Wolfe, and J. Hajjar. 2014. Non-EOS Root Causes of EOS-Like Damage, in *Proceedings of the EOS/ESD Symposium*, EOS/ESD Association, Rome, NY, pp. 7A.2.1–7A.2.6.
2. R. Wong, S. Wen, R. Fung, and P. Le. 2013. System Level EOS Case Studies Not Due to Excessive Voltages, in *Proceedings of the EOS/ESD Symposium*, EOS/ESD Association, Rome, NY, pp. 7A.2.1–7A.2.6.
3. International Electrotechnical Commission. 2012. IEC 61000-4-4, *Electromagnetic Compatibility—Part 4-4: Testing and Measurement Techniques—Electrical Fast Transient/Burst Immunity Test*, International Electrotechnical Commission, Geneva, Switzerland.
4. International Electrotechnical Commission. 2014. IEC 61000-4-5, *Electromagnetic Compatibility—Part 4-5: Testing and Measurement Techniques—Surge Immunity Test*, International Electrotechnical Commission, Geneva, Switzerland.
5. EOS/ESD Association and JEDEC Solid State Technology Association. *ANSI/ESDA/JEDEC JS-001-2014, Human Body Model, Component Level*, EOS/ESD Association, Rome, NY.
6. JEDEC Solid State Technology Association. JESD22-C101F, *Field-Induced Charged Device Model: Test Method for Electrostatic-Discharge-Withstand Thresholds of Microelectronic Components* JEDEC Solid State Technology Association, Arlington, VA.
7. M. Chaine and M. Etherton. 2012. Charged Device Model Phenomena and Circuit Design, *Predictive Circuit Simulation of Charged Device Model ESD Events, Tutorial at 2012 ESD Symposium* EOS/ESD Association, Rome, NY.
8. EOS/ESD Association and JEDEC Solid State Technology Association. ANSI/ESDA/JEDEC JS-002-2014, *Charged Device Model—Device Level*. EOS/ESD Association, Rome, NY, and JEDEC Solid State Technology Association, Arlington, VA.
9. EOS/ESD Association. *ANSI/ESD/STM 5.2 Machine Model, Component Level*.
10. JEDEC document JEP172. 2014. *Discontinuing use of the machine model for device ESD qualification*. JEDEC Solid State Technology Association, Arlington, VA.
11. R.R. Troutman. 1986. *Latchup in CMOS Technology*, Kluwer, Norwell, MA.
12. JEDEC Solid State Technology Association, Arlington, VA. September 2010. JESD78D, *IC Latch-Up Test*.
13. ESD Association. 2013. ESD TR5.4-04-13, *Transient Latch-Up Testing, Working Group 5.4*.
14. A. Ochoa, Jr., F.W. Sexton, T.F. Wrobel, G.L. Hash, and R.J. Sokel. December 1983. Snap-Back: A Stable Regenerative Breakdown Mode of MOS Devices, *IEEE Trans. Nucl. Sci*, Institute of Electrical and Electronic Engineers, Piscataway, NJ, Vol. NS-30, No. 6, pp. 4127–4130.
15. A. Olney, B. Gifford, J. Guravage, and A. Righter. 2003. Real-World Charged Board Model (CBM) Failures, in *Proceedings of the EOS/ESD Symposium*, pp. 1–10.

5 Design of Power-Rail ESD Clamp Circuits with Gate-Leakage Consideration in Nanoscale CMOS Technology

Ming-Dou Ker and Chih-Ting Yeh

CONTENTS

5.1 INTRODUCTION

Electrostatic discharge (ESD) phenomenon is a charge flow when two objects with different voltage potentials reach contact. Such ESD events can cause serious damage to the integrated circuit (IC) products, during assembly, testing, and manufacturing. To protect the IC products with the required ESD specifications, typically such as 2 kV in human body model (HBM) [1] and 200 V in machine model (MM) [2], the whole-chip ESD protection scheme formed with the power-rail ESD clamp circuit had been often used in the modern IC products [3]. As shown in Figure 5.1, the power-rail ESD clamp circuit is a vital element for ESD protection under different ESD stress modes. The ESD stress modes include V_{dd}-to-V_{ss} (or V_{ss}-to-V_{dd}) ESD stress between the rails, as well as the positive-to-V_{ss} (PS) mode, negative-to-V_{ss} (NS) mode, positive-to-V_{dd} (PD) mode, and negative-to-V_{dd} (ND) mode, from input/output (I/O) to V_{dd}/V_{ss}. Therefore, the power-rail ESD clamp circuit must provide low-impedance discharging path under ESD events but keep in off-state with standby leakage current as low as possible under normal circuit operation conditions.

In advanced nanoscale CMOS technology, there are two commonly used processes provided from foundry for some specified purposes. They are low-power (LP) and general-purpose (GP) processes. LP process is used for LP product with a 1.2 V core design and 2.5 V or 3.3 V I/O option. Because LP process is developed for LP product, there is basically no serious gate leakage issue. Therefore, a large-sized MOSFET drawn in the layout style of big field-effect transistor (BigFET) is usually adopted as the ESD clamp device in the power-rail ESD clamp circuit.

GP process provides higher performance transistors for high-speed or high-frequency applications with 1 V core design and 2.5 V I/O option. In GP process, the thickness of gate-oxide layer is thinner than that in LP process (or with a lower threshold voltage, V_{th}) to gain higher driving current. However, the thinner gate oxide impacts

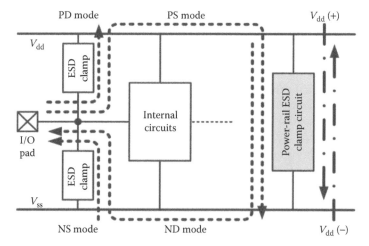

FIGURE 5.1 Typical whole-chip ESD protection scheme with the power-rail ESD clamp circuit under different ESD stress conditions.

TABLE 5.1
Gate-Leakage Current of MOS Capacitor in GP Processes

Generation	MOS Type	Effective Oxide Thickness (nm)	Gate Leakage Current at 1 V ($W/L = 1$ μm/1 μm) (nA)
90 nm (w/o HKMG)	nMOS	~2.3	~11
	pMOS	~2.5	~3
65 nm (w/o HKMG)	nMOS	~2.0	~140
	pMOS	~2.2	~80
45 nm (w/o HKMG)	nMOS	~1.9	~260
	pMOS	~2.1	~95
28 nm (w/ HKMG)	nMOS	~1.35	~123
	pMOS	~1.4	~42

seriously on the ESD protection circuits because of the intolerable gate leakage and the lower breakdown voltage. A comparison of gate-leakage issue on MOS capacitor ($W/L = 1$ μm/1 μm) among different CMOS technologies is shown in Table 5.1. In the 28 nm technology node, the structure of high-k/metal gate (HKMG) [4] has been adopted to reduce the gate-leakage current issue and to continuously shrink the effective oxide thickness (EOT). However, the large-sized MOSFET in those advanced processes was obviously inadequate to meet the requirement of low standby leakage current. Hence, silicon-controlled rectifier (SCR) without poly-gate structure has been adopted as the main ESD clamp device in the power-rail ESD clamp circuits.

Recently, some circuit techniques have been developed to reduce the gate-leakage current and layout area of the power-rail ESD clamp circuits. In this chapter, those different circuit techniques are reviewed and discussed. The layout area, ESD robustness, and standby leakage current among those designs are compared.

5.2 PROCESS TECHNIQUES USED TO IMPLEMENT THE POWER-RAIL ESD CLAMP CIRCUIT

5.2.1 TRADITIONAL RC-BASED POWER-RAIL ESD CLAMP CIRCUIT WITH THICK GATE OXIDE

The traditional RC-based power-rail ESD clamp circuit was widely used to protect the core circuits [3], as shown in Figure 5.2. Using the thick gate oxide can directly avoid the gate-leakage issue. The RC-based ESD-transient detection circuit commands the ESD clamp NMOS transistor to turn on under ESD stress condition and to turn off under normal circuit operation condition. The turn-on time of the ESD clamp NMOS transistor can be adjusted by the RC time constant of the RC-based ESD-transient detection circuit to meet the half-energy discharging time of the HBM ESD event [1]. To meet the aforementioned requirements, the RC time constant of the RC-based ESD-transient detection circuit is typically designed about 0.1–1 μs to

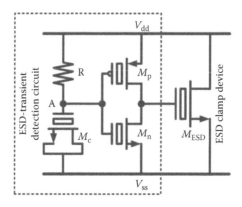

FIGURE 5.2   Traditional *RC*-based power-rail ESD clamp circuit with thick gate oxide and large-sized NMOS transistor as ESD clamp device. (Data from Ker, M.-D., *IEEE Trans. Electron Devices*, 46, 1, 173–183, 1999.)

achieve the desired operations. As the *RC* networks in the microsecond range are somehow large, it would occupy a significant fraction of the layout area.

5.2.2   Using the High-*k*/Metal Gate Structure

Gate dielectric has been one of the major challenges for technology scaling. With leakage and reliability constraints, high gate leakage of silicon dioxide has limited further scaling of gate dielectric thickness particularly for high performance applications. Besides, it is imperative to replace poly-Si with metal gate to eliminate poly-depletion. To continue EOT scaling, high-*k*/metal gate CMOS technology resumes gate dielectric scaling and is a solution to the gate leakage. However, the ideal high-*k*/metal gate technology needs some requirements of good performance integrity of higher *k* value, low leakage current, low threshold voltage, high mobility, and thermal stability for ion-implant doping activation [5–8].

5.2.3   Using the Parasitic Capacitance between Metal Layer (MOM Capacitor)

Metal oxide metal (MOM) capacitors have been commonly used in IC design because the MOM capacitor has higher linearity, higher quality factor (Q), small temperature variation, and almost no leakage current [9]. When the dimensions keep shrinking in advanced CMOS technologies, the capacitance density of MOM is increased significantly and MOM capacitor will not occupy large chip area. Therefore, the MOM capacitor used in the ESD-transient detection circuit can solve the leakage issue.

The power-rail ESD clamp circuit with MOM capacitor is shown in Figure 5.3 [10], which consists of the ESD-transient detection circuit with MOM capacitor and the p-type triggered SCR as the ESD clamp device. Without the thin gate-oxide structure, SCR has very low leakage current under normal circuit operating condition. Besides,

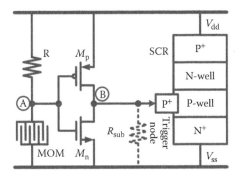

FIGURE 5.3 Power-rail ESD clamp circuit with MOM capacitor. (Data from Chiu, P.-Y. and Ker, M.-D., "Design of low-leakage power-rail ESD clamp circuit with MOM capacitor and STSCR in a 65 nm CMOS process," in *Proceedings of the International Conference on Integrated Circuit Design & Technology*, IEEE, 2011.)

SCR had been proven to have the highest ESD robustness under the smallest device size [11]. Moreover, SCR can be safely used without latch-up danger in advanced CMOS technologies of low supply voltage.

Under normal power-on condition, the voltage level at node A can follow up the voltage level at V_{dd} power line to keep the M_p off. Simultaneously, the M_n is turned on because its gate terminal is connected to node A. As the result, no trigger current is injected into SCR, and SCR can be kept off. Figure 5.4 shows the simulated transient waveforms with a rise time of 0.1 ms. With the power supply voltage of 1 V, the simulated overall leakage current of the power-rail ESD clamp circuit is only about 307 nA at 25°C.

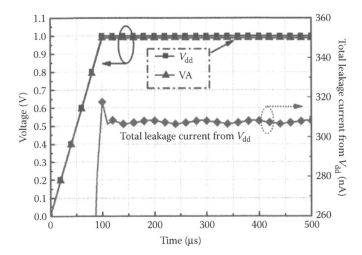

FIGURE 5.4 Simulated transient waveforms of the ESD-transient detection circuit with MOM capacitor under normal power-on transition.

5.3   CIRCUIT TECHNIQUES USED TO IMPLEMENT THE POWER-RAIL ESD CLAMP CIRCUIT

5.3.1   Feedback-Enhanced Triggering Technique

The power-rail ESD clamp circuit with feedback-enhanced triggering was depicted in Figure 5.5 [12]. A fast positive-going voltage pulse on the V_{dd} power rail causes node A to rise instantaneously along with the V_{dd} potential. The elevation of node A causes M_{ESD} to be turned on, and ESD current can be discharged. Once the potential of node C has been raised to the voltage level above the threshold voltage of M_{nf}, M_{nf} begins to conduct. Current conduction in transistor M_{nf} further pulls the potential of node B toward ground, which further enhances current conduction in transistor M_{p2}, which then pulls the potential of node C closer to that of the V_{dd} power line. This completes a feedback loop to latch M_{ESD} into a conductive state.

Once M_{ESD} has been latched into a conductive state, the time constant of the RC circuit is now free to time out. The duration of this time constant can be significantly shorter than the ESD event, which translates into an RC network with greatly reduced physical area. However, the transistors of the feedback loop and the ESD clamp device in this design need some modifications to reduce the gate leakage.

5.3.2   Cascaded PMOS-Feedback Technique

The power-rail ESD clamp circuit with cascaded PMOS-feedback technique was shown in Figure 5.6 [13]. This circuit uses a small capacitor in the RC timer, relative to the traditional RC-based design. One of the immediate advantages is that MOSFET-size rationing is not critical for this circuit. Of course, the cascaded PMOS should not be so small as to affect the switching speed of INV2 and subsequently INV3.

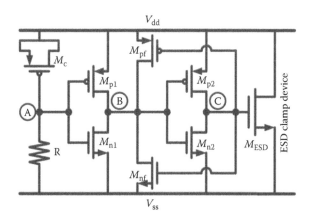

FIGURE 5.5 Power-rail ESD clamp circuit with feedback-enhanced triggering and ESD clamp NMOS transistor. (Data from Smith, J.C. and Boselli, G., "A MOSFET power supply clamp with feedback enhanced triggering for ESD protection in advanced CMOS technologies," in *Proceedings of the EOS/ESD Symposium*, IEEE, 2003, pp. 8–16.)

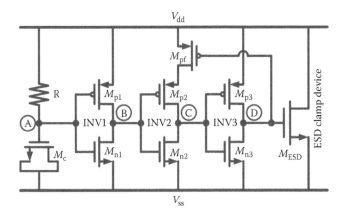

FIGURE 5.6 Power-rail ESD clamp circuit with cascaded PMOS-feedback technique and ESD clamp NMOS transistor. (Data from Li, J. et al., "A compact, timed-shutoff, MOSFET-based power clamp for on-chip ESD protection," in *Proceedings of the EOS/ESD Symposium*, IEEE, 2004, pp. 273–279.)

Upon initiation of a positive ESD event between V_{dd} and V_{ss}, the voltage at node A stays low for a time period set by the RC time constant. This low voltage at node A causes node C at 0 V. The feedback PMOS M_{pf} has no effect at this time because M_{p2} is turned off. The low voltage at node C causes the voltage at node D to be pulled up to V_{dd}. Thus, M_{ESD} is fully conducting within three inverter delays when the ESD stress is initiated. Once the voltage at node A rises above the switching threshold of INV1, M_{n2} is turned off and M_{p2} is turned on. However, as the voltage at node D is V_{dd}, M_{pf} is turned off and node C remains in the low state. As long as the voltage at node C is less than the threshold voltage of M_{n3}, the gate voltage of M_{ESD} will not be perturbed and M_{ESD} stays in turned-on conduction beyond the time constant of the RC network. Similar to the design in Section 5.3.1, all transistors in the ESD detection circuit and the ESD clamp device require more attention to reduce the gate leakage.

5.3.3 Reducing the Gate Area of the MOS Capacitor

A power-rail ESD clamp circuit with smaller capacitance that adopts the capacitance-coupling mechanism has been shown in Figure 5.7 [14]. The cascade NMOS transistors (M_{nc1} and M_{nc2}) operated in the saturation region are used as a large resistor and combined with the smaller capacitor to construct a capacitance-coupling network.

Under ESD stress condition, the potential of node A will be synchronously elevated toward a positive voltage potential by capacitance coupling of the smaller capacitor. Then, the gate terminal of the ESD clamp NMOS transistor will be promptly charged toward the positive voltage potential. Under normal circuit operation condition, the potential of node A will actually be kept at V_{ss} through the high-resistance path of the cascade NMOS transistors. Therefore, the ESD clamp NMOS transistor will be kept at the off-state under normal circuit operation condition. For reducing the total standby leakage current of this design in nanoscale CMOS

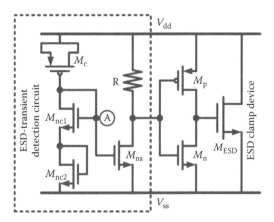

FIGURE 5.7 Power-rail ESD clamp circuit with smaller capacitance in ESD-transient detection circuit. (Data from Chen, S.-H. and Ker, M.-D., *IEEE Trans. Circuits Syst. [II]*, 56, 5, 359–363, 2009.)

technology, the device sizes of inverter can be shrunk, and the ESD clamp device can be replaced by SCR.

5.3.4 REDUCING THE VOLTAGE DROP ACROSS THE MOS CAPACITOR (I)

The equations of the gate-direct-tunneling current from BSIM4 MOSFET model [15] indicate that the leakage current through the MOS capacitor can be reduced by reducing the voltage across it. Based on this concept, the power-rail ESD clamp circuit with feedback-control inverter to overcome the gate leakage issue is shown in Figure 5.8 [16]. In the ESD-transient detection circuit, the *RC*-based ESD-transient detection circuit and the feedback-control inverter are combined together, and the MOS capacitor M_{cap} is connected between the nodes A and B. Because M_{cap} is not directly connected to V_{ss}, no direct leakage path is conducted through M_{cap} to the ground under normal circuit operating condition. Without the thin gate oxide, the SCR used as the main ESD clamp device is also free to the gate leakage issue as compared with a large-sized MOSFET.

With a slow rise time of the normal power-on transition, the voltage level at node A will be able to follow up the voltage level at V_{dd} power line to keep M_{p1} off. The parasitic p-substrate resistor R_{sub} in SCR can pull node C to V_{ss}. The M_{p3} would also be turned on to drive the node B to V_{dd}. With the voltage of V_{dd} at node B, M_{p2} can be fully turned off. In addition, M_{n1} is turned on because its gate terminal is connected to node B. Obviously, there is no voltage drop across M_{cap}, and no circuit leakage path exists in the ESD-transient detection circuit. Without a voltage drop across M_{cap} under normal circuit operating condition, M_{cap} can be realized with a large device size without suffering the leakage current. As nodes A and B are charged to V_{dd}, M_{p1} and M_{p2} can be fully turned off during the normal power-on transition. Therefore, no trigger current is injected into the SCR, and the SCR can be kept off under normal circuit operating condition.

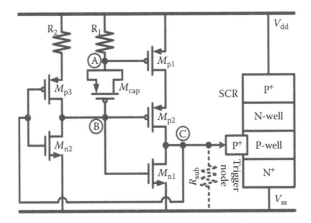

FIGURE 5.8 Power-rail ESD clamp circuit with feedback-control inverter. (Data from Ker, M.-D. and Chiu, P.-Y., *IEEE Trans. Device Mat. Rel.*, 11, 3, 474–483, 2011.)

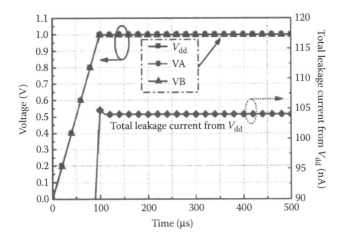

FIGURE 5.9 Simulated transient waveforms on the node voltages in the ESD-transient detection circuit with feedback-control inverter under normal power-on transition.

Figure 5.9 shows the simulated transient waveforms of the ESD-transient detection circuit under the normal power-on transition with a rise time of 0.1 ms. With the power supply voltage of 1 V, the overall simulated leakage current of the ESD-transient detection circuit is only about 104 nA at 25°C.

5.3.5 Reducing the Voltage Drop across the MOS Capacitor (II)

The power-rail ESD clamp circuit with the consideration of the gate current is shown in Figure 5.10 [17]. The SCR device is used as the main ESD clamp device. Utilizing the gate current to bias the ESD-transient detection circuit and to reduce the voltage difference across the gates of the MOS capacitors, the gate leakage current through the MOS capacitor under the normal circuit operating condition can be further reduced.

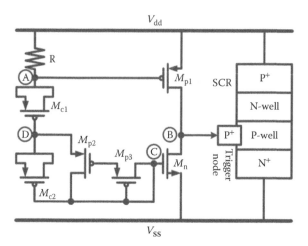

FIGURE 5.10 Power-rail ESD clamp circuit with utilization of gate current. (Data from Wang, C.-T. and Ker, M.-D., *IEEE J. Solid-St. Circuits*, 44, 3, 956–964, 2009.)

Therefore, the total leakage current resulted from the MOS capacitor in the ESD-transient detection circuit can be well controlled and minimized.

In the ESD-transient detection circuit, M_{p1} is used to generate the triggering current into the trigger node of the SCR during the ESD stress event, but M_{p1} is kept off under the normal circuit operating condition. The M_n is used to keep the voltage level at the trigger node (node B in Figure 5.10) at V_{ss}, so the SCR is guaranteed to be turned off during the normal circuit operating condition. The RC time constant from R, M_{c1}, M_{c2}, and the parasitic gate capacitance of M_n is designed around the order of microsecond to distinguish ESD stress event from the normal power-on condition. The diode-connected M_{p2} and M_{p3} are acted as a start-up circuit with initial gate-to-bulk current from V_{dd} into the ESD-transient detection circuit, and in turn to conduct some gate current of M_{c1} to bias nodes C and D. After that, the voltage level at node D will be biased to reduce the voltage difference across M_{c1} and to minimize the gate-leakage current through the MOS capacitors.

Figure 5.11 shows the simulated voltage waveforms on the nodes of the ESD-transient detection circuit and the gate current through the MOS capacitor M_{c1} under the normal power-on condition with a rise time of 1 ms and V_{dd} of 1 V (V_{ss} of 0 V). The gate voltage of M_{p1} is biased at 1 V through the resistor R with a low gate current (~23 nA) of MOS capacitor M_{c1}, so that M_{p1} can be kept off and no trigger current is generated into the SCR device. In addition, node C is biased at 0.45 V to turn on M_n, which in turn keeps the trigger node of SCR grounded.

5.3.6 CAPACITOR-LESS DESIGN OF POWER-RAIL ESD CLAMP CIRCUIT

The capacitor-less design of power-rail ESD clamp circuit is illustrated in Figure 5.12 [18]. The power-rail ESD clamp circuit consists of the ESD-transient detection circuit with feedback technique, which is realized by two transistors (M_n and M_p) and two resistors (R_n and R_p), and the ESD clamp NMOS transistor (M_{ESD}) drawn in BigFET

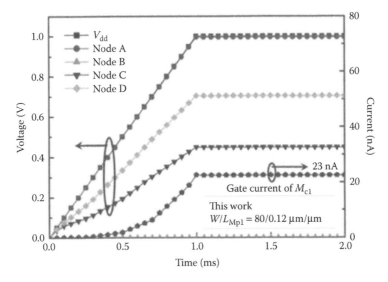

FIGURE 5.11 Simulated voltage on the nodes and the gate current flow through the MOS capacitor M_{c1} of the ESD-transient detection circuit with utilization of gate current under normal power-on transition.

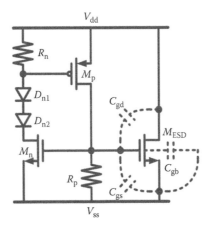

FIGURE 5.12 Capacitor-less power-rail ESD clamp circuit with diode string in the ESD-transient detection circuit and ESD clamp NMOS transistor. (Data from Yeh, C.-T. and Ker, M.-D., *IEEE J. Solid-St. Circuits*, 45, 11, 2476–2486, 2010.)

layout style. The gate terminal of M_{ESD} is linked to the output of the ESD-transient detection circuit. The ESD-transient detection circuit with positive feedback mechanism is constructed by a cascade structure (R_n with M_n and M_p with R_p), which can command M_{ESD} at on- or off-state. To overcome the transient-induced latch-on issue, the ESD-transient detection circuit is added with diode string to adjust its holding voltage.

Because the ESD clamp NMOS transistor is drawn in BigFET layout style without silicide blocking, large C_{gd}, C_{gs}, and C_{gb} parasitic capacitances essentially exist in the ESD clamp NMOS transistor. Sufficiently utilizing these parasitic capacitances with the R_p to realize capacitance-coupling mechanism, no additional capacitor is needed in this design. Under ESD stress condition, the M_n immediately starts the ESD-transient detection circuit when the voltage of node A is elevated by capacitance coupling. When the subthreshold current of the M_n can produce enough voltage drop on R_n to further turn on the M_p, the voltage at node A would be elevated quickly to the voltage level at V_{dd} because the ESD-transient detection circuit is turned on. Consequently, the M_{ESD} is turned on by the ESD-transient detection circuit with positive feedback mechanism. Although the leakage current can be reduced because of no actual capacitor device in ESD-transient detection circuit, the ESD clamp device drawn in BigFET layout style still contributes large gate-leakage current. Therefore, some modifications of this design are required in nanoscale CMOS process as discussed in the following.

The modified power-rail ESD clamp circuit with p-type triggered SCR as the main ESD clamp device is shown in Figure 5.13 [19]. The ESD-transient detection circuit is designed with considerations of the gate-leakage current and the gate-oxide reliability. In Figure 5.13, the M_p is used to generate the trigger current into the trigger node C of the p-type triggered SCR during the ESD stress event. Under the normal circuit operating condition, the M_p is kept off, and the trigger node is kept at V_{ss} through the parasitic p-substrate resistor R_{sub}. Therefore, the p-type triggered SCR device is turned off during the normal circuit operating condition.

Because of the lack of parasitic capacitor of BigFET, the RC-based ESD-transient detection mechanism is realized by the R_n and the junction capacitance of the reverse-biased diode D_c. The reverse-biased diode D_c used as the capacitor can be free from the gate-leakage current issue. The inserted diodes, D_n and D_p, in the ESD-transient detection circuit are used to reduce the voltage differences across the gate oxide of

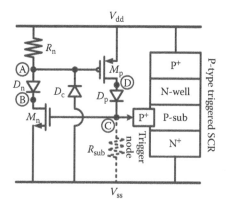

FIGURE 5.13 Power-rail ESD clamp circuit with positive feedback. (Data from Yeh, C.-T. and Ker, M.-D., *IEEE Trans. Electron Devices*, 59, 10, 2626–2634, 2012.)

the transistors M_p and M_n in the ESD-transient detection circuit. Therefore, the gate-leakage current and gate-oxide reliability of M_p and M_n can be well controlled to minimize the total standby leakage current.

Under the normal circuit operation condition with V_{dd} of 1 V and grounded V_{ss}, the gate voltage of M_p is biased at 1 V through the resistor R_n. The gate voltage of M_n is biased at 0 V simultaneously through the parasitic p-substrate resistor R_{sub}. Because M_p is kept off, no trigger current is generated into the trigger node of SCR. By inserting the diodes, D_p and D_n, in the ESD detection circuit, the voltages at nodes B and D can be clamped to the desired higher or lower voltage levels. Therefore, the drain-to-gate and drain-to-source voltages of M_p and M_n can be far less than 1 V to further reduce the standby leakage current.

The simulated voltage waveforms and the leakage current of the ESD-transient detection circuit during the normal power-on transition are shown in Figure 5.14, where V_{dd} is raising from 0 to 1 V with a rise time of 1 ms. In Figure 5.14, the voltage differences across the gate-to-drain, gate-to-source, and drain-to-source terminals of all transistors in the ESD-transient detection circuit are only about 0.5 V. The simulated leakage current of the ESD-transient detection circuit is around 13.9 nA for the p-type triggered design.

5.3.7 RESISTOR-LESS DESIGN OF POWER-RAIL ESD CLAMP CIRCUIT

The resistor-less design of power-rail ESD clamp circuit is shown in Figure 5.15 with the p-type triggered SCR as the main ESD clamp device [20]. The ESD-transient detection circuit is also designed with considerations of the gate-leakage current and the gate-oxide reliability. The RC-based ESD-transient detection mechanism is realized by the equivalent resistors (R_{gs} and R_{gd}) of M_p and the junction capacitance

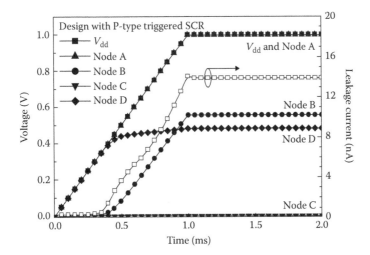

FIGURE 5.14 Simulated voltage waveforms on the nodes and the leakage current of the ESD-transient detection circuit with positive feedback under the normal power-on transition.

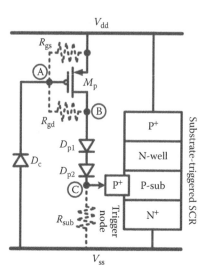

FIGURE 5.15 Resistor-less design of power-rail ESD clamp circuit with diode string in the ESD-transient detection circuit. (Data from Yeh, C.-T. and Ker, M.-D., *IEEE Trans. Electron Devices*, 59, 12, 3456–3463, 2012.)

of the reverse-biased diode D_c, which can distinguish the ESD stress event from the normal power-on condition. By using the gate-leakage current of the M_p, the induced equivalent resistors can be a part of ESD-transient detection mechanism to achieve the resistor-less design. In Figure 5.15, the M_p is mainly used to generate the trigger current into the trigger node C of SCR during the ESD stress event. Comparing to the thin gate oxide of MOS capacitor in the traditional RC circuit, the diode D_c used as capacitor to realize the RC time constant can be free from the gate-leakage current issue. The inserted diodes, D_{p1} and D_{p2}, in the ESD-transient detection circuit are used to reduce the voltage differences across the gate oxide of M_p. Therefore, the total leakage current and gate-oxide reliability of M_p can be safely relieved.

Under the normal circuit operation condition, the gate voltage of M_p is biased at V_{dd} through the resistors R_{gs} and R_{gd} induced by the gate-leakage current. The cathode of D_{p2} is simultaneously biased at V_{ss} through the parasitic p-substrate resistor R_{sub}. Because M_p is kept off, no trigger current is generated into the trigger node of SCR. Inserting two diodes (D_{p1} and D_{p2}) in the ESD-transient detection circuit can raise up the voltage of node B at the voltage level near to V_{dd}. Therefore, all terminals of M_p are almost at the same voltage level of V_{dd} to reduce its gate-leakage current.

The simulated voltage waveforms and the leakage current of the resistor-less ESD-transient detection circuit during the normal power-on transition are shown in Figure 5.16. In Figure 5.16, the voltage of node A is successfully charged to the voltage level of V_{dd} because of the gate-leakage current. Therefore, the M_p is completely turned off, and the simulated standby leakage current of the ESD-transient detection circuit is only 1.53 nA.

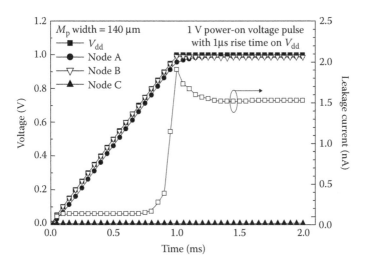

FIGURE 5.16 Simulated voltage waveforms on the nodes and the leakage current of the resistor-less ESD-transient detection circuit under the normal power-on transition.

5.3.8 DIODE-STRING ESD DETECTION CIRCUIT

The power-rail ESD clamp circuit designed with diode-string ESD detection is shown in Figure 5.17 with the p-type triggered SCR as the main ESD clamp device [21]. This design was implemented with a diode string and a resistor to detect the ESD events by the high-voltage level instead of the fast rise time.

Under normal circuit operation, the V_{dd} operating voltage is lower than the diode string threshold voltage. Therefore, there is no current flowing through R, and M_p is kept off. Adding a voltage drop by using a diode D_o between M_p drain and the SCR

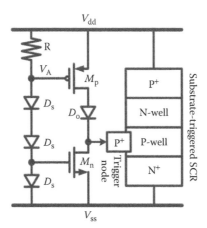

FIGURE 5.17 Power-rail ESD clamp circuit with diode-string ESD detection. (Data from Altolaguirre, F.A. and Ker, M.-D., *IEEE Trans. Electron Devices*, 60, 10, 3500–3507, 2013.)

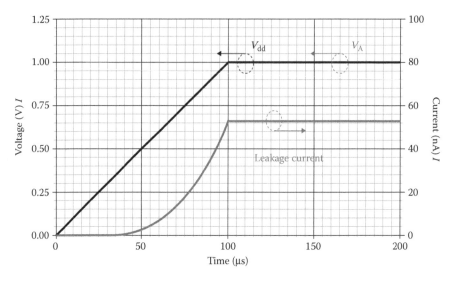

FIGURE 5.18 Simulated voltage waveforms on the nodes and the leakage current of the diode-string ESD detection circuit under the normal power-on transition.

trigger point (V_{TRIG}) would effectively reduce the leakage current from M_p. Under a positive-to-V_{ss} ESD stress, the diode string starts to conduct some current when the V_{dd} voltage overpasses the diode string threshold voltage. That causes a voltage drop across R, thus turning M_p on to trigger the SCR.

The simulated results of this design with diode-string ESD detection during the normal power-on transition are shown in Figure 5.18. In Figure 5.18, the voltage (V_A) of node A is successfully charged to the voltage level of V_{dd}. Therefore, the M_p is completely turned off, and the simulated standby leakage current is only 52 nA.

5.4   DISCUSSION AND COMPARISON

The comparison among various power-rail ESD clamp circuits is summarized in Table 5.2. Some evaluated parameters are explained as following [22].

5.4.1   STANDBY LEAKAGE CURRENT

For the standby leakage current, the designs of 2.A and 2.C are moderate. For the design of 2.C, although the MOS capacitor is replaced by MOM capacitor, the voltage difference across the transistors of inverter is not reduced sufficiently. By carefully considering the voltage difference across the gate oxide, the standby leakage current of the other designs (3.D–3.H) can be greatly reduced. Especially for the design of 3.G, the measured standby leakage current is only a few nanoamperes because all terminals of the MOSFET are biased at the same voltage level of V_{dd}.

TABLE 5.2

Comparison among Power-Rail ESD Clamp Circuits

ESD Protection Design	Measured Standby Leakage Current at Normal Circuit Operation Voltage	HBM ESD Robustness	Area Efficiency	Design Complexity	Mis-Triggered
2.A—Traditional RC-Based	Moderate	Better	Poor	Low	No
2.C—MOM Capacitor	Moderate	Good	Poor	Low	No
3.A—Feedback-Enhanced Triggering	Good	Good	Moderate	Moderate	Yes
3.B—Cascaded pMOS Feedback	Moderate	Moderate	Moderate	Moderate	Yes
3.C—Smaller Capacitance	Moderate	Better	Moderate	Moderate	No
3.D—Feedback Control Inverter	Good	Moderate	Moderate	Moderate	No
3.E—Utilization of Gate Current	Good	Better	Better	Moderate	No
3.F—Positive Feedback	Better	Good	Good	Low	No
3.G—Resistor-Less Design	Excellent	Good	Better	Low	No
3.H—Diode-String Design	Excellent	Good	Better	Low	No

5.4.2 HBM ESD ROBUSTNESS

In the design of 3.B, the HBM ESD robustness is moderate for over 3 kV. With over 2000 µm channel width of ESD clamp device, the HBM ESD robustness of design 3.A is good to be over 5 kV. For the designs of 2.A and 3.C, the HBM ESD robustness is better due to over 8 kV.

In designs of 2.C, 3.F, and 3.G, the HBM ESD robustness is good for over 4 and 5 kV, respectively. With 120 µm SCR width in the design of 3.D, the HBM ESD robustness is moderate due to over 8 kV. However, the HBM ESD level of design 3.E (3.H) is better to be 7 kV (6.5 kV) with only 45 µm (40 µm) SCR width.

5.4.3 AREA EFFICIENCY AND DESIGN COMPLEXITY

The area efficiency of traditional RC-based design (2.A) is poor because the RC time constant is typically designed about 0.1–1 µs. It would consume large layout area to implement resistor and capacitor, but the design complexity of traditional RC-based design is low. The layout area of design 2.C with MOM capacitor is poor because the ESD-transient detection circuit is based on traditional RC-based design and still consumes large layout area to implement resistor and MOM capacitor.

Some previous designs (3.A and 3.B) with feedback mechanism were presented to reduce the *RC* time constant and layout area. However, the reduction of layout area is limited because additional feedback circuits are required in the ESD-transient detection circuit, which increase the level of design complexity from low to moderate.

For the design of 3.C, the area efficiency and the design complexity are both moderate, as compared to traditional *RC*-based design. The layout area of design 3.D with feedback-control inverter is moderate because the resistances are reduced to consume smaller layout area. The layout areas of the other designs (3.E–3.H) are good, even better because the device dimensions in ESD-transient detection circuit are greatly reduced. However, the designs of 3.D and 3.E require more devices to reduce the voltage difference across the gate oxide, which increase the level of design complexity as moderate.

5.4.4 Mis-Triggered

Some previous studies [23,24] have demonstrated that the power-rail ESD clamp circuits with *RC*-based ESD-transient detection circuits were easily mis-triggered or into the latch-on state under the fast power-on condition. Therefore, this issue exists in the designs of 3.A and 3.B.

The designs of 3.C and 3.D can be safely applied to fast power-on condition without mis-triggered issue. According to the circuit structure, the designs of 3.F–3.H can also avoid those issues by adjusting the number of diode in diode string.

5.5 SUMMARY

A comprehensive overview on the design of power-rail ESD clamp circuits in the nanoscale CMOS technology has been presented. Some process and circuit techniques used in the ESD-transient detection circuits were adopted to perform better turn-on behavior, lower standby leakage current, and higher efficiency of layout area. Generally, SCR is adopted as main ESD clamp device in the power-rail ESD clamp circuits because of no poly-gate structure. With considerations of the gate-leakage current and the gate-oxide reliability, the total standby leakage current in some advanced designs has been successfully reduced to the order of a few nanoamperes. Continuously, the power-rail ESD clamp circuit will still be an important design task for on-chip ESD protection as the process is further scaling down.

REFERENCES

1. *Electrostatic Discharge Sensitivity Testing—Human Body Model (HBM)—Component Level*, ESD Association Standard, Rome, NY, 2001, Test Method ESD STM5.1.
2. *Electrostatic Discharge Sensitivity Testing—Machine Model (MM)—Component Level*, ESD Association Standard, Rome, NY, 1999, Test Method ESD STM5.2.
3. M.-D. Ker, "Whole-chip ESD protection design with efficient VDD-to-VSS ESD clamp circuits for submicron CMOS VLSI," *IEEE Trans. Electron Devices*, vol. 46, no. 1, pp. 173–183, January 1999.

4. W.C. Shen, C.Y. Mei, Y.-D. Chih, S.-S. Sheu, M.-J. Tsai, Y.-C. King, and C.J. Lin, "High-K metal gate contact RRAM (CRRAM) in pure 28 nm CMOS logic process," in *International Electron Devices Meeting Technical Digest*, IEEE, 2012, pp. 745–748.

5. Y.T. Hou et al., "High performance tantalum carbide metal gate stacks for nMOSFET application," in *International Electron Devices Meeting Technical Digest*, IEEE, 2005.

6. C.H. Wu et al., "High temperature stable [Ir$_3$Si-TaN]/HfLaON CMOS with large work-function difference," in *International Electron Devices Meeting Technical Digest*, IEEE, 2006.

7. H.T. Huang et al., "45 nm high-k/metal-gate CMOS technology for GPU/NPU applications with highest PFET performance," in *International Electron Devices Meeting Technical Digest*, IEEE, 2007, pp. 285–288.

8. C.H. Diaz et al., "32 nm gate-first high-k/metal-gate technology for high performance low power applications," in *International Electron Devices Meeting Technical Digest*, IEEE, 2008.

9. H. Samavati, A. Hajimiri, A. Shahani, G. Nasserbakht, and T. Lee, "Fractal capacitors," *IEEE J. Solid-St. Circ.*, vol. 33, no. 12, pp. 2035–2041, December 1998.

10. P.-Y. Chiu and M.-D. Ker, "Design of low-leakage power-rail ESD clamp circuit with MOM capacitor and STSCR in a 65 nm CMOS process," in *Proceedings of the International Conference on Integrated Circuit Design & Technology*, IEEE, 2011.

11. M.-D. Ker and K.-C. Hsu, "Overview of on-chip electrostatic discharge protection design with SCR-based devices in CMOS integrated circuits," *IEEE Trans. Device Mat. Rel.*, vol. 5, no. 2, pp. 235–249, June 2005.

12. J.C. Smith and G. Boselli, "A MOSFET power supply clamp with feedback enhanced triggering for ESD protection in advanced CMOS technologies," in *Proceedings of the EOS/ESD Symposium*, IEEE, 2003, pp. 8–16.

13. J. Li, R. Gauthier, and E. Rosenbaum, "A compact, timed-shutoff, MOSFET-based power clamp for on-chip ESD protection," in *Proceedings of the EOS/ESD Symposium*, IEEE, 2004, pp. 273–279.

14. S.-H. Chen and M.-D. Ker, "Area-efficient ESD-transient detection circuit with smaller capacitance for on-chip power-rail ESD protection in CMOS ICs," *IEEE Trans. Circuits Syst. II*, vol. 56, no. 5, pp. 359–363, May 2009.

15. BSIM Model, *Berkeley Short-Channel IGFET Model*. [Online]. Available: http://www-device.eecs.berkeley.edu/bsim.

16. M.-D. Ker and P.-Y. Chiu, "New low-leakage power-rail ESD clamp circuit in a 65 nm low-voltage CMOS process," *IEEE Trans. Device Mat. Rel.*, vol. 11, no. 3, pp. 474–483, September 2011.

17. C.-T. Wang and M.-D. Ker, "Design of power-rail ESD clamp circuit with ultra-low standby leakage current in nanoscale CMOS technology," *IEEE J. Solid-St. Circuits*, vol. 44, no. 3, pp. 956–964, March 2009.

18. C.-T. Yeh and M.-D. Ker, "Capacitor-less design of power-rail ESD clamp circuit with adjustable holding voltage for on-chip ESD protection," *IEEE J. Solid-St. Circuits*, vol. 45, no. 11, pp. 2476–2486, November 2010.

19. C.-T. Yeh and M.-D. Ker, "Power-rail ESD clamp circuit with ultralow standby leakage current and high area efficiency in nanometer CMOS technology," *IEEE Trans. Electron Devices*, vol. 59, no. 10, pp. 2626–2634, October 2012.

20. C.-T. Yeh and M.-D. Ker, "Resistor-less design of power-rail ESD clamp circuit in nanoscale CMOS technology," *IEEE Trans. Electron Devices*, vol. 59, no. 12, pp. 3456–3463, December 2012.

21. F.A. Altolaguirre and M.-D. Ker, "Power-rail ESD clamp circuit with diode-string ESD detection to overcome the gate leakage current in a 40 nm CMOS process," *IEEE Trans. Electron Devices*, vol. 60, no. 10, pp. 3500–3507, October 2013.

22. M.-D. Ker and C.-T. Yeh, "On the design of power-rail ESD clamp circuits with gate leakage consideration in nanoscale CMOS technology," *IEEE Trans. Device Mat. Rel.*, vol. 14, no. 1, pp. 536–544, March 2014.

23. C.-C. Yen and M.-D. Ker, "The effect of IEC-like fast transients on *RC*-triggered ESD power clamps," *IEEE Trans. Electron Devices*, vol. 56, no. 6, pp. 1204–1210, June 2009.

24. C.-T. Yeh and M.-D. Ker, "High area-efficient ESD clamp circuit with equivalent *RC*-based detection mechanism in a 65 nm CMOS process," *IEEE Trans. Electron Devices*, vol. 60, no. 3, pp. 1011–1018, March 2013.

6 ESD Protection in Automotive Integrated Circuit Applications

Javier A. Salcedo and Jean-Jacques Hajjar

CONTENTS

Automotive electronics is increasingly becoming a fundamental component as well as an ever-appealing feature in modern automotive technology. Electronics have been embedded in the automobile's core mechanical system in applications such as adaptive suspension, throttle, and engine control as well as braking systems, to name a few. This has been necessary for improved driver safety. The trend toward more efficient and "greener" cars has also contributed to greater integration of electronics components in the automobile, with the advent of hybrid and electrical vehicles. Additionally, the differentiating features of modern car technology appeal to the comfort and convenience of the driver. Information technology such as multimedia, speech recognition, and navigation assistance has found application in most modern automobiles. The challenge is to combine these new features and functionalities with the requisite quality and robustness.

Premium automobiles contain over 7000 semiconductor components [1,2]. Managing this level of electronic integration into a mechanical system is a daunting task especially when the reliability and quality of these components have to meet very stringent robustness requirement because of their use in automotive systems—a highly safety critical application. Furthermore, electronics play a major role in enabling growing sophistication and complex signal processing in automotive and industrial applications. To create a safer operation environment and greener, lighter, and more comfortable automobiles of the future, technology advancements in data converters (analog to digital converters [ADCs] and digital to analog converters [DACs]), amplifiers, MEMS (micro-electromechanical systems), sensor interface ASICs (application-specific integrated circuits), digital isolators, processors, and high-frequency technologies are instrumental in enabling cost-effective and reliable system designs. Electronic devices have been integrated over time to replace mechanical devices and systems, subsequently improving accuracy and sparking innovation for rollover and stability control, crash safety, radar and vision driver assistance, HEV (hybrid electric vehicle)/EV (electric vehicle) power management, fuel injection, transmission control, and infotainment, among other sensor and signal processing applications. Figure 6.1 depicts a high-level representation of electronics control in a generic automobile. These different components include combination of electronic sensors, activators, drives, and control units, working in synergy for executing safety and comfort functions in the car environment.

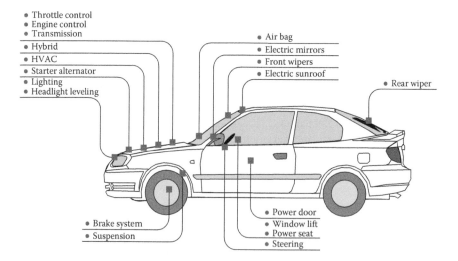

FIGURE 6.1   Schematic representation of automotive electronics.

Table 6.1 lists typical electronic application examples in automotive with the corresponding comments on specifications. These applications range from automotive power standard products and application-specific standard products (ASSPs) to highly integrated customized ASICs. Along with the ASICs, integrated sensor systems would include different MEMS and mixed-signal processing capability. By enabling highly integrated module solutions for increasingly complex electromechanical systems control, new functionality for greater precision and overall better performance is achieved. On the other hand, reliability of these electronic components at the automotive operating conditions is one of the fundamental considerations, as it raises the bar on the process safe operating area (SOA) capability and controls at different stages of the design flow.

Figure 6.2 shows a graphical representation of an advanced driver assistance system (ADAS). Among the attractive monitoring options, radar and vision complementary signal processing is currently considered [2,3]. In this scheme, the ADAS technologies provide the driver with essential information to prevent accidents. The objective is the automation of difficult or repetitive tasks to improve car safety. Among the different applications, in this particular example, are adaptive cruise control, adaptive breaking control, adaptive light control, automotive parking, and collision avoidance. Additionally, the concept of integrated automotive safety systems (ISSs) is another example that merges ideas such as pre-crash sensing, anticipatory crash sensing, X-by-wire systems (e.g., steering, braking, throttle, and suspension that de-couple the actuation from the mechanical input provided by the driver), advanced safety interiors, integrated vehicle electronics systems, data networks, and mobile multimedia (telematics). These technologies result in an improved driving experience and better overall road safety.

For these electronic applications, electromechanical sensors generate signals that need to be transmitted to the engine electronics control unit (ECU) to be interpreted. The ECU responds to the sensor signal and activates the proper function, such as the deployment of the air bag, the switching of the ABS (antilock braking system) valves, or the supply of power to the electric engine. These functions are controlled by a microcontroller and mixed-signal power modules [2,3]. Beyond the microcontroller, the automotive electronics are manufactured either with HV-CMOS and/or smart power technologies such as bipolar-CMOS-DMOS (BCD) or with common-drain IC processes [4–10].

IC manufacturing process technologies for automotive and other IC applications operating in harsh environments must meet stringent requirements, in particular if they are located outside the passenger compartment or directly interfacing with other electronics components, for example, the ECU, via harness cabling. These requirements can include wide operating voltage (e.g., −80 V to 80 V), operation at more extreme temperature (e.g., −40°C to 300°C), and the capability of safely handling relatively high-energy disturbances. Among these high-energy disturbances, EMI (electromagnetic interference) and ESD (electrostatic discharge) are examples of random and unpredictable high-energy stress conditions that can occur at any time during system integration and field operation [2–8].

This chapter reviews the design considerations and the solution approaches for such circuit applications operating in harsh environment. The emphasis is driven

TABLE 6.1

Automotive Electronics Applications

Application	Example Functions	System Elements and Considerations
Safety	Air bag systems	MEMS satellite sensor accelerometer. Example communication includes PSI5-compliant (125–148 kB/s). ASICs are often implemented in different high-voltage CMOS and BCD processes.
	Electric power steering	iGMR (integrated giant magneto resistive), AMR (anisotropic magneto resistive) sensors. Applications can include different sensing components, e.g., AMR position sensing, accelerometers, gyroscopes. ASICs are often implemented in different high-voltage CMOS and BCD processes.
	Rollover and stability control	Gyroscope, low-G accelerometer. ASICs are often implemented in different high-voltage CMOS and BCD processes.
	Pressure sensor	Signal conditioning IC, including its communication interface, connecting to the automotive cabling system. ASICs are often implemented in different high-voltage CMOS and BCD processes.
	ADAS (radar)	Short-range radar (24 GHz) and short/long-range radar (77 GHz). Integrated at the passenger compartment close to the ECU (see Figure 6.2).
	ADAS (vision)	System integrated at the passenger compartment close to the ECU (see Figure 6.2).
	Tire pressure monitor	It ensures correct tire pressure by local pressure monitoring and wireless RF short-range transmission to automotive ECU.
Body applications	Door, lights, windows, interior light, control module, start/stop systems	Fully integrated multi-IC embedded control modules and discrete devices drivers.
Infotainment	Analog and embedded processing systems for video, audio, and in-car multimedia	Video and audio systems, signal conditioning amplifiers. These devices are typically placed in the passenger compartment.
Powertrain	Optimization of gear ratio and gear shift	Transmission control modules (embedded systems including microprocessor and high-voltage discrete devices).
Hybrid and electric vehicles control	Lithium ion battery safety monitor, control system, and high-power electronics drivers	Example components can include high-voltage BCD battery monitors and control, along with discrete high-power drivers.

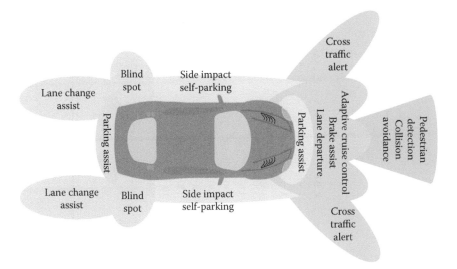

FIGURE 6.2   Schematic representation of an ADAS.

toward automotive interface applications. The first section describes the standards for EMI and ESD considered for this type of applications. The impact of ESD and EMI on the technology development would also be addressed with particular attention to reference device design window margin and protection device development. Guidelines on technology characterization and quality controls for enabling this type of circuit applications would be discussed with particular consideration on the trade-offs encountered in process technologies and cost.

6.1   CONDUCTIVE OVERSTRESS ROBUSTNESS CHARACTERIZATION

Unlike conventional mobile communication or other consumer applications, automotive electronics are subject to a wider range of stress conditions. Table 6.2 lists commonly used standards to emulate stress conditions for device-level ESD, transient system-level ESD, and EMI immunity for automotive applications. The stress conditions defined by the ISO 10605 and IEC 61000-4-2 standards can be applied to the pins and system housing, but the specific combination can vary among electronics systems and different circuit applications. For instance, air discharge is often applied to housing plastic discharge points, as it would be the typical case of automotive satellite or pressure sensors. This stress condition, however, is also often required to be applied directly at metallic conductive pins for interface applications [10–17].

The longer-duration EMI (>500 ns in duration) events emulated by the different ISO standards in Table 6.2 are caused, for example, by the switching of high-power systems such as actuators, relays, and motors. This resulting energy spreads over the network. Another source of interference is caused by coupling of electromagnetic

TABLE 6.2
Summary of Reference Conductive Disturbance Test Standards

Standard	Application	V (V)	Duration (10%–90%)	# Pulses	R_i (Ω)	C (pF)	I_{peak} (A)
AEC-Q100: JS-001-2012: HBM	Component	2 k	~150 ns	1	1.5 k	100	~1.5
AEC-Q100: JESD22-101: FICDM	Component	750	~0.7 ns	1	1	Package-dependent	~5–20
ISO 10605/IEC 61000-4-2	System	~8 k-contact discharge	120 ns	10	330	150	~30
ISO 10605/IEC 61000-4-2	System	~15 k-air discharge	120 ns	10	330	150	~30
ISO 10605	System	~8 k-contact discharge	150 ns	10	330	330	~30
ISO 10605	System	~15 k-air discharge	150 ns	10	330	330	~30
ISO 10605	System	~8 k-contact discharge	360 ns	10	2 k	150	~30
ISO 10605	System	8 k	1 μs	10	2 k	330	~30
ISO 7637: 1	System	~150	~2 ms	500	10	–	~15
ISO 7637: 2a	System	~112	~50 μs	500	2–10	–	~56
ISO 7637: 2b	System	~10	~0.2–2 s	10	0.05	–	~200
ISO 7637: 3a	System	~220	~150 ns	3.6E6 (1 h)	50	–	~4.5
ISO 7637: 3b	System	~150	~150 ns	3.6E6 (1 h)	50	–	~3
ISO 16750-2 (load dump)	System	~101	~40–400 ms	10	2	–	~50

radiation to the device or system and can last for a few nanoseconds up to several hundred milliseconds (see Table 6.1). One of the most serious stress conditions, the ISO 16750-2 (load dump), is a result of high energy injected into the system. This can happen, for instance, when the car battery is suddenly disconnected while the alternator is still charging. Comparing the EMI stress duration with the shorter-duration (<500 ns) ESD events in Table 6.2, the on-chip ESD protection elements itself is not intended to dissipate the energy of the ISO 7637 or ISO 16750-2 stress conditions. In principle, the ISO energy is expected to be dissipated by external discrete devices (e.g., transient voltage suppressors [TVS]) on the PCB (printed circuit board), designed to clamping the voltage to safe values for the IC component.

Besides the aforementioned standards emulating different automotive environmental stress conditions, direct RF power injection (DPI) is an electromagnetic susceptibility (EMS) test standard defined by the International Electrotechnical Commission [18]. It forms part of the EMI test standards under the heading of electromagnetic compatibility (EMC). The DPI stress test is part of the EMS tests, among bulk current injection (BCI) [19] and ESD tests.

The communication between decentralized ECUs is typically provided by controller area network (CAN), local interface network (LIN), or FlexRay bus systems [20,21], thereby making the interface circuits to be directly exposed to these stress conditions. As automotive electronics becomes an even larger segment of the semiconductor industry, the ICs must be reliable in these harsh environments, forcing the semiconductor manufacturers and foundries to develop more robust process technologies.

6.2 PROCESS TECHNOLOGY CONSIDERATIONS FOR AUTOMOTIVE AND INDUSTRIAL APPLICATIONS

Table 6.3 compares process technology often considered for automotive applications [5,7,22–34]. Notice that different applications would consider a large variety of process technologies depending on the design constraints. The robustness of a technology for circuit operation under high-stress conditions, such as automotive, can include evaluation of the core devices and protection devices transient SOA (TSOA) [35–37], as it aids defining in a systematic way the conditions at which the circuit can safely operate. As an example, the standard 12 V interface system voltage specification typically uses a nominal technology voltage of 60 V. In this particular case, a significant margin has to be added to the system voltage to take into account destructive stress conditions previously discussed. The latter must not activate the internal circuit junction's breakdowns. In particular, high-energy stress conditions from previous discussion, [13–18] must be clamped off-chip, usually by external capacitors or TVS. The maximum clamping voltage of these elements is typically below 50 V. This limit defines the minimum voltage capability of the IC technology.

To illustrate example test setups, equivalent circuit, and simulated waveforms of those stress conditions discussed in the previous section, Figure 6.3a and b shows reference test setup for ESD GUN testing, following ISO 10605 and IEC 61000-4-2 standards convention [11–12]. In the case of Figure 6.3a, the test configuration emulates a floating ground device under test (DUT), often obtained in battery-power electric

TABLE 6.3

Typical Process Technologies Used for Automotive IC Applications

Process Technology	Low-Voltage Circuit	High-Voltage Devices	Substrate	General Considerations
Multi-gate voltage CMOS	CMOS	HV CMOS and DMOS	Varies; typically lightly doped P-type substrate	For relatively high-voltage applications, typically >30 V, this process integrates EPI and buried layer. The use of multiple gate oxide allows for full gate drive applications, simplifying circuit architecture implementation. The multiple gates add cost to the process.
High-voltage-tolerant mixed-signal CMOS	CMOS	DMOS, EDMOS	Typically lightly doped P-type substrate	Mixed-signal CMOS process enhanced to allow for high-voltage devices, with optimum use of lightly doped deep well isolation, and without the need of EPI and buried layers. They are considered as low-cost options for implementing automotive applications, typically targeting <30 V operation.
High-voltage SOI	CMOS	DMOS	Silicon on insulator allows for deep trench	It is a high-cost option, but traditionally widely adopted for automotive applications because of robustness consideration and elimination of substrate parasitic. The drawback of SOI is its limited capability to build large power stages. Due to the buried oxide, a weaker thermal conductivity to the substrate is a limitation of this process.
High voltage with deep trench	CMOS	DMOS	Varies; typically highly doped substrate	A deep trench isolation (DTI) high-voltage process is considered a mid-range between the conventional shallow trench (STI) isolation BCD process and a fully isolated DTI SOI process.
High-frequency CMOS	Advanced sub 65 nm/28 nm CMOS	EDMOS, CMOS	Typically highly doped	Advanced process technologies considered for processor units, auto audio, and high-frequency applications.
High-frequency BiCMOS	SiGe BiCMOS process	CMOS, bipolar transistors	NA	High-frequency technology considered for applications such as automotive radar.
High-voltage BCD	CMOS	DMOS	Varies	It is a lower-cost technology when compared with an SOI process. The suppression of parasitic substrate bipolar transistors and thyristors within a BCD technology is a major challenge, which becomes even more important with further shrinkage.

components [11]. On the other hand, Figure 6.3b emulates the condition in which the ground of the DUT is connected directly to the main ground of the system, for instance, the chassis or car body. In each case, the DUT would be subject to stress conditions that better emulate the final operating environment, which changes between application categories and whether the IC is inside the passenger compartment or

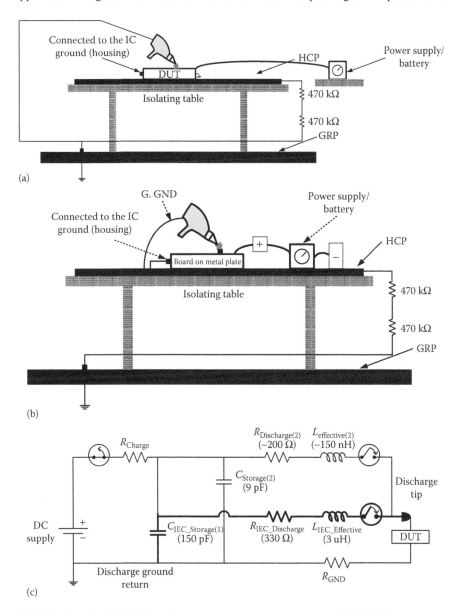

FIGURE 6.3 (a) IEC 61000-4-2 testing setup for floating system, (b) IEC 61000-4-2 testing setup for grounded system, and (c) IEC 61000-4-2 equivalent schematic for circuit simulation.

(*Continued*)

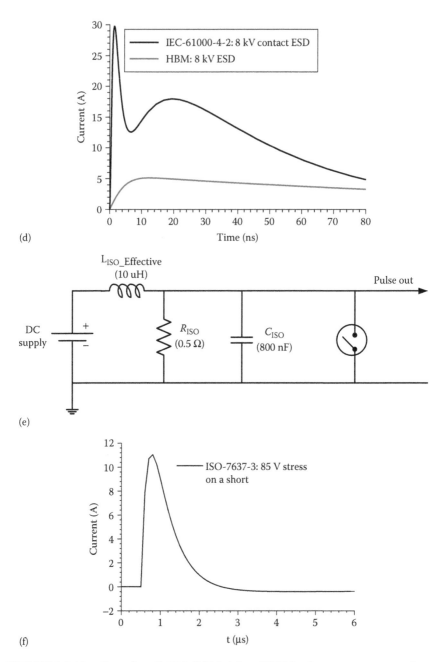

FIGURE 6.3 (Continued) (d) IEC 61000-4-2 and HBM reference current waveforms for 8000 V, (e) ISO 7637-3 equivalent schematic for circuit simulation, and (f) simulated ISO 7637-3 current waveform for 85 V. (*Continued*)

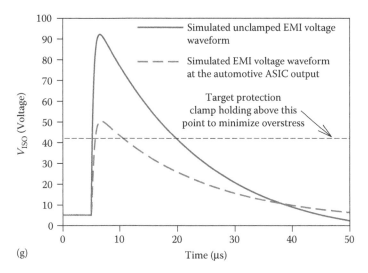

(g)

FIGURE 6.3 (Continued) (g) example of simulated voltage waveforms unclamped and at the IC I/Os for an 85 V EMI target (ISO 7637-3 pulse 2).

close to the engine. Figure 6.3c and d shows a simplified equivalent circuit for the IEC 61000-4-2 for circuit simulation and the corresponding 8,000 V stress waveform following this standard. For reference, an 8,000 V HBM current waveform has been added in light gray to the figure to emphasize the meaningful difference in the level of current between these stress conditions. Notice in Figure 6.3c the use of the simulation switches between the different networks, a method to emulate and fine-tune the double-peak characteristics in the IEC 61000-4-2 current waveform.

Figure 6.3e shows a simplified circuit to simulate a reference ISO 7637-3 current pulse, shown in Figure 6.3f. This EMI current pulse would have a longer duration, often requiring the use of external components to absorb a greater portion of this stress. This condition is shown in Figure 6.3g, in which an unclamped voltage level simulated at the interface can reach over 90 V at the IC pin level, but with addition of a decoupling capacitor, the peak voltage at the automotive ASIC is less than 55 V in this example. This voltage-level constraints the technology that can be used for automotive applications subject to these conditions at the interface pins.

The on-chip protection devices, on the other hand, are expected to get activated above the operating conditions of the circuit but below the point at which the circuit's internal devices show irreversible failure, which typically can be identified as the breakdown voltage of the technology core devices. This voltage range in the ESD design requirements is called *ESD design window* or *ESD design margin* [36]. The trigger voltage determines the beginning of ESD protection circuitry operation. Designing a lower trigger voltage protection clamp is beneficial in protecting the internal circuit devices, but as the trigger gets too close to the operating voltage, it may also induce ESD structure mistriggering by normal circuit operation events. A higher trigger voltage on the other hand delays the ESD triggering, impacting the design capability to protect the internal devices. Balancing this trade-offs is a key task when assessing design architectures and device topologies suitable for these applications.

6.3 COMMUNICATION INTERFACE FOR AUTOMOTIVE APPLICATIONS

Reliable communication between components in the car environment plays an important role in the automotive electrification. In the case of automotive applications, LIN, CAN, and FlexRay protocols are the most commonly used bus systems. LIN is used for lower-cost, low data-rate functions, while CAN allows for faster speed communication. FlexRay is the latest developed network and selected for high-speed data rates and safety-critical applications. These different communication interfaces are implemented using different circuit design techniques and increasingly being connected directly to the car wiring harness. As a result, they share the requirement of meeting high level of robustness at the communication interface pins, which include passing most of the standard stress conditions summarized in Table 6.2. This section will review the design considerations and example applications focusing on reference LIN and CAN implementation examples for assessment of the design trade-offs.

6.4 AUTOMOTIVE LIN

LIN is a low-cost serial network and multiplexing protocol used for communication between components in vehicles. Automotive ECUs that implement the LIN communicate with each other over a one-wire data bus [1]. In today's car networking topologies, microcontrollers with either UART (universal asynchronous receiver/ transmitter) capability or dedicated LIN hardware are used. The microcontroller generates all needed LIN data and is connected to the LIN network via a LIN transceiver. Current main uses combine the low-cost efficiency of LIN and sensors to create automotive electronics networks.

6.5 EXAMPLE OF LIN DESIGN IN A BCDMOS TECHNOLOGY: ESD DESIGN OPTIMIZATION

Figure 6.4 shows the fundamental components that can be considered for the design of the LIN interface in a 60 V process. This type of interface can be used in products such as intelligent battery sensor, capacitance convertor, or sensor signal conditioner ASICs. The protection components can be co-designed along with the interface circuit to achieve robustness design targets in a more compact footprint.

The low side of the LIN interface can be formed using a DMOS device (MN_0) in series with a custom high-voltage blocking P–N junction (D_0). This high-voltage blocking junction allows the output signal swing to go below the LIN ground and is formed by the base–emitter junction of a substrate PNP (SPNP). This latter can be particularly sensitive to ESD damage during the high-stress levels obtained at the LIN interface pin.

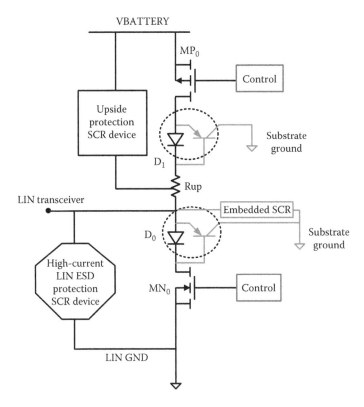

FIGURE 6.4 Simplified schematic of a LIN interface fundamental circuit components. Device symbols shown in gray are parasitic devices.

The formation of this SPNP is a main design constraint for enabling this type of circuit applications in a low-cost junction-isolated process. To address this design challenge, the device can be modified to incorporate embedded protection [38]. This avoids the SPNP from triggering and getting damaged when subjected to the high-stress conditions. To prevent damage in the SPNP, as the base–collector voltage increases beyond approximately 55 V, the protection should start to conduct current. This has two advantages: (1) It protects the weak B–C junction of the SPNP, and (2) it helps prevent the NLDMOS from being damaged during a snapback event.

6.6 INTERFACE PINS REFERENCE DESIGN

6.6.1 VBATTERY ESD DESIGN

The VBATTERY pin can be protected to ground by a bidirectional clamp [39–41], with a high-voltage reverse protection diode in parallel. This reverse protection diode provides an ESD path during HBM-type ESD events, when the external series diode is not present. The breakdown voltage of the bidirectional clamp can be adjusted by junction engineering without adding extra cost to the process [25].

The protection clamp can be optimized in this case to provide a trigger voltage in the range of 45 V. As a result and under extreme conditions, for instance during ISO 7637-2 and IEC 61000-4-2 stress conditions, when many amps can flow into the protection clamp, the voltage across the reverse protection diode remains below its reverse breakdown voltage. The reverse protection diode can be designed in this case to handle relatively high transient current, for example, >10 A, thereby giving it extra headroom to sink excess current resulting from the limitation of the external protection.

6.6.2   LIN PIN ESD DESIGN REFERENCE

6.6.2.1   LIN Protection Clamp

The bidirectional clamp on the LIN pin can be an optimized version of the VBATTERY pin clamp structure. It can use the same triggering mechanism and spacing. It can also be adjusted accordingly to achieve different turn-on characteristics [25]. The VBATTERY clamp can have a holding voltage of around 45 V. High holding voltages are required on clamps that will be tied to the car battery, as they will provide enough current to destroy the device if it does not turn off. In the application, the LIN pin can have a minimum resistance to VBATTERY of 1 kΩ. Therefore, it does not need to have a high holding voltage to guarantee its turn-off. This allows a trade-off between turn-on time and holding voltage that is not available on the VBATTERY pin.

6.6.2.2   LIN Reverse Blocking Diode

In conventional junction-isolated BCD processes, all diodes are in fact bipolar junction transistor (BJT) devices. In the regular high-voltage N-well to high-voltage P-well (HVPW–HVNW) diode, the diode is the base–emitter junction of an SPNP, and the collector terminal must always be considered [38]. During an ESD event, the voltage on the anode (emitter) of the SPNP can increase to >100 V. The cathode (base) of this device is connected to an NLDMOS device. The NLDMOS device starts to conduct, and a significant amount of current starts flowing in the emitter–base junction. As there is also a high voltage across the base–collector junction, the SPNP can easily get damaged. To prevent this damage, Figure 6.5 shows a cross section of a reference diode with embedded protection [38]. When the base–collector voltage increases, the protection formation between the different terminals of the device enhances its robustness by discharging the overstress current before damage occur at relatively low stress level.

Figure 6.6 shows the characterization of the blocking diode with embedded protection when subject to wafer-level-like (IEC 61000-4-2) stress condition. The plots depict the resulting characteristics for the negative stress pulse, which closely correlates with the required level for +7.5 and −6.5 kV IEC 61000-4-2 ESD.

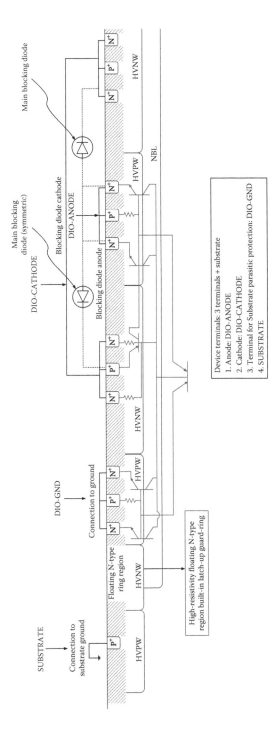

FIGURE 6.5 Reference cross section for blocking junction with built-in SCR device. (Data from Clarke, D. et al., "Junction-isolated blocking voltage devices with integrated protection structures and method of forming the same," US Patent 8,796,729, August 2014.)

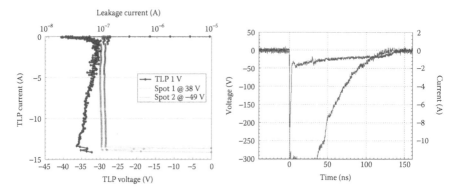

FIGURE 6.6 Quasi-static (left) and transient (right) measurement for wafer-level IEC 61000-4-2-like pulse in the LIN pin diode stressed with negative voltage. (Data from Clarke, D. et al., "Junction-isolated blocking voltage devices with integrated protection structures and method of forming the same," US Patent 8,796,729, August 2014.)

6.7 HIGH-VOLTAGE GUARD-RING CONSIDERATIONS

Figure 6.7 shows an example of a high-voltage guard-ring array adopted for separating regions subject to high stress [4,38]. Under these conditions, the activation of large protection devices can inject relatively large current to the substrate, resulting in unintended parasitic devices activation. This makes this type of application particularly complex when it comes to designing for latch-up robustness, as conventional latch-up design rules used for standard applications would need to be adjusted to enhance the guard-ring robustness at the higher temperature, higher levels of current obtained in LIN automotive applications.

The guard-ring example in Figure 6.7, from left to right, starts with the HVPW guard ring; centered inside the HVPW, there is a P+ active. Next to the HVPW is a HVNW guard ring formed similarly by an N+ centered inside the HVNW. This guard ring is biased to reduce risk of triggering a parasitic bipolar or latch-up path.

6.8 CONTROLLER AREA NETWORK

CAN bus is a vehicle bus standard designed to allow microcontrollers and devices to communicate with each other within a vehicle without a host computer. It is a message-based protocol, designed specifically for automotive applications but now

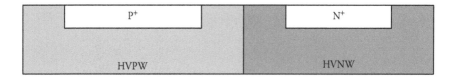

FIGURE 6.7 Cross-sectional view of simplified latch-up guard-ring formation.

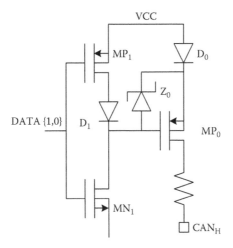

FIGURE 6.8 High-level view of CAN_H schematic.

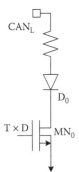

FIGURE 6.9 High-level view of CAN_L schematic.

also used in other areas such as aerospace, maritime, industrial automation, and medical equipment. CAN has become a multi-master serial bus standard for connecting ECUs, also known as *nodes*. Two or more nodes are required on the CAN to communicate [21]. The complexity of the node can range from a simple I/O device up to an embedded computer with a CAN interface and sophisticated software. Figures 6.8 and 6.9 show example schematics for the CAN_H and CAN_L, respectively. The protection clamp for these pins follow a similar concept to the one discussed in the previous section in connection with the LIN interface.

6.9 EMERGING AUTOMOTIVE INTERFACE APPLICATIONS

6.9.1 AUTOMOTIVE AUDIO SERIAL BUS LOW-VOLTAGE DIFFERENTIAL SIGNALING TRANSCEIVER ARCHITECTURE

As part of the automotive environment signal communication chain, the innovative adoption of automotive audio bus (A2B) transceivers for infotainment applications enables a cost-effective method of multiplexing and transmitting audio data from

multiple microphone or microphone array endpoints to the head unit or specialized ECU. In particular, the A2B topology allows for a flexible open daisy chain
and tree transceiver architecture with single master and multiple slave devices [42].
Figure 6.10 shows a simplified diagram of the transceiver interface pins and the
corresponding dual-polarity ground-reference protection devices at the interface
pins (squares). These pins are typically subject to direct stress during circuit operation, requiring them to meet the relatively high level of system-level ESD and EMI
immunity standards. The corresponding high-level schematic for the transceiver
pins implementation is shown in Figure 6.11. Cascaded MOS devices for the low and
high side are used in this architecture to extend the design protection window for
the primary interface protection clamp, enabling a higher breakdown for positive or
negative stress conditions at the interface pin to ground.

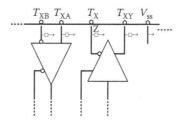

FIGURE 6.10 Diagram of transceiver pins including dual-polarity ESD protection to common ground (V_{ss}).

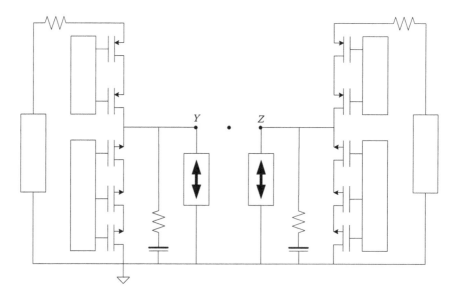

FIGURE 6.11 A2B LVDS transceiver schematic view.

6.10   ESD PROTECTION DEVICE DESIGN AND MEASUREMENTS

Figure 6.12 shows the cross-sectional view of the ±8 V blocking voltage dual-tub (N-type buried layer [NBL] and deep P-well [DPW]) ESD protection device designed for protection at the A2B transceiver communication pins [43]. The dual-polarity blocking junction is defined in this device between the shallow N-well (ShNW) (N⁺) and the shallow P-well (ShPW) junction formation underneath the RPO (resist protection oxide) regions. This blocking junction formation defines the bidirectional blocking voltage in the device, which can be design-adjusted to operate at different bidirectional blocking voltages within the flexibility provided by the various N- and P-type implants combination in the BCD process. For instance, in a typical 40–60 V BCD process, removing the N⁺ region would increase the blocking voltage to about ±16 V breakdown between the ShNW and the ShPW. It would be even higher breakdown by allowing HVNW to HVPW (DPW) junction formation (~30–45 V).

Figure 6.13 shows the equivalent schematic representation of the device in Figure 6.12. The main conduction path during protection activation is defined by the SCR-like formation by devices Q_{p1}, Q_{n1}, and Q_{p2}. Focusing the attention on the floating base NPN Q_{n1}, the emitter–base junction formation aids in defining the bidirectional blocking junction, and this device in combination with the PNP's Q_{p1} and Q_{p2} is activated to form a high conductivity and high current–handling capability

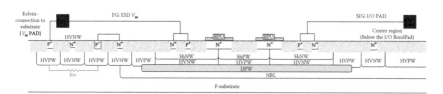

FIGURE 6.12   Cross-sectional view of ±8 V blocking voltage dual-tub (NBL and DPW) ESD protection device for protection at the transceiver communication pins in A2B. (Data from Salcedo, J.A., "Low voltage protection devices for precision transceivers and methods of forming the same," US Patent 8,610,251, December 2013.)

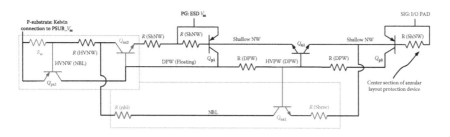

FIGURE 6.13   Equivalent schematic for ESD protection device in Figure 6.18. (Data from Salcedo, J.A., "Low voltage protection devices for precision transceivers and methods of forming the same," US Patent 8,610,251, December 2013.)

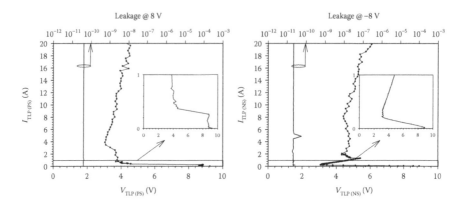

FIGURE 6.14   Absolute value of ESD device bidirectional TLP measurement results.

discharge path during stress conditions. The bipolar devices, Q_{px1}, Q_{nx1}, and Q_{nx2}, are parasitic devices formed between each terminal of the protection device and the substrate. As the terminal PG is typically connected to the same substrate potential, Q_{nx2}'s emitter and collector are effectively at the same potential. The combination of devices Q_{px1} and Q_{nx1}, however, does require more attention as this SCR-like formation between the SIG I/O PAD terminal and the substrate can negatively impact the device functionality by activating an undesirable conduction path during stress or low breakdown to substrate.

Figure 6.14 shows measurement results from the example described earlier for the positive and negative stress conditions from the terminal SIG to terminals PG and substrate connected to ground. The device shows high current–handling capability (>20 A) and trigger voltage in the ±8 V target range. The trigger voltage at the 8 V level is required to protect the low-voltage core circuit devices directly connected to the interface pins in Figure 6.11. Besides the TLP measurement results, the device was characterized for robustness during an ESD stress, passing over 10,000 V as indicated by the IEC 61000-4-2 specifications. The device consists of a low blocking voltage bidirectional clamp adapted to protect isolated low-voltage interface pins. This protection device also includes a high-voltage isolation guard-ring array (see Figure 6.12), for trigger control and to minimize substrate injection at stress conditions occurring while the system is powered. The fast turn-on of the protection clamp in this case is critical to guaranteeing that the stress signal stays within voltage ranges safe for the circuit. The latter can be extracted from gate oxide and conventional core devices in the process [36,37,44].

6.11 AUTOMOTIVE QUALITY SYSTEM

The automotive market demands the highest standard in terms of reliability and quality, delivery timeliness, as well as cost of its materials. This requirement is applied across all its suppliers. To address this in a structured process from concept to mass production is necessary to meet these demands. Advanced quality and control planning techniques have been devised to manage the complexity of product development [45,46]. The rigor in this approach stems from it categorizing the tasks into separate phases, which include planning, product design and development, process design and development, product and process validation, and mass production. An in-depth description of these planning steps is beyond the scope of this chapter. A few key elements and processes will be mentioned to provide insight into the scope of the overall tasks.

6.12 MANUFACTURING PROCESS TECHNOLOGY SELECTION

The choice of IC manufacturing process technology is critical as it will help address key design circuit challenges. The electrical stress limits and also fault coverage conditions will determine the applicability of the process to the design. Although there are design techniques to overcome some of the shortcomings in the process, a balance between process robustness and circuit complexity needs to be established. The latter will have repercussion on the timeliness of the product delivery as well as cost.

6.13 DESIGN FOR MANUFACTURABILITY

This design stage is critical as it is an important differentiator from the competitions and consequently defines the commercial marketability of the product. Successful designs are not based only on electrical performance but based on the manufacturability of the product as well. If this is not taken into consideration, it will lead to design iteration with loss of manufacturing time and marketing opportunity through longer time to market. Organizations have adopted different practices for ensuring proper design for manufacturing, and some are listed here.

6.13.1 DESIGN HIERARCHY AND REUSE

A hierarchical design approach is typically the norm for complex circuits. As bottoms-up approach while it may leverage the manufacturing process technology capability with circuit design acumen, it may have limited applicability when integrated within the overall system. Block-level design must comprehend the requirement of the system prior to detailed design. Moreover, previously designed blocks that meet these requirements and have been proven in mass production would be ideal candidate for reuse. The latter would contribute toward reducing project's schedule risk.

6.13.2 DESIGN ROBUSTNESS

Spurious and unintentional electrical stresses that can either interfere with the performance of the product or cause unrecoverable failures in it have to be considered. Design techniques for making the product robust to such interferences are to be implemented as part of the design and tested through various industry standardized test developed for this purpose.

6.13.3 Manufacturing Variation

A robust design that is immune to manufacturing variability is desirable and contributes to the manufacturability of the product. There are various computer-aided design (CAD) tools that have been developed to help product designers achieving this objective. Circuit simulation tools are available to probe the sensitivity of a design to manufacturing process variability as well as device mismatch within a circuit. Additionally, design rules for reliability, yield, and assembly are well-established guidelines on mature processes that can be checked in the design through automated CAD tools.

6.13.4 Process Design and Development

An effective manufacturing flow is required post design to ensure that controls are in place to deliver a quality product. In addition, it has to be developed to meet customer's needs and requirements. A comprehensive assessment of the functionality is usually undertaken for all datasheet parameters.

The control levels may broadly be divided into four categories, namely

1. Wafer probe test methods
2. Final test method
3. Statistical bin and yield limits
4. Statistical defect control

For additional controls for a product's early stage of production, a safe launch plan is initiated. The latter's objective is to deliver a defect free part for, typically, a safety critical end application. Additional testing and controls are usually required to achieve the intended goal.

Finally, a key process in the completion of the design is the production product approval. Its purpose is to ensure that the design specification comply with the customer requirements. The engineering records and specification should be well understood by the supplier, and the manufacturing process should be able to produce the product consistently.

REFERENCES

1. C. Ching-Yao, "Trends in crash detection and occupant restraint technology," *Proceedings of the IEEE*, vol. 95, no. 2, pp. 388–396, February 2007.
2. U. Abelein, H. Lochner, D. Hahn, S. Straube, "Complexity, quality and robustness—the challenges of tomorrow's automotive electronics," in *EDAA*, IEEE, 2012, http://publica.fraunhofer.de/documents/N-219704.html. [Accessed May 24, 2015.]
3. P. Peti, R. Obermaisser, F. Tagliabo, A. Marino, S. Cerchio, "An integrated architecture for future car generations," in *IEEE International Symposium on Object-Oriented Real-Time Distributed Computing*, pp. 2–13, May 2005, http://www.ieee.org/index.html. [Accessed May 24, 2015.]
4. J.A. Salcedo, J.-J. Hajjar, S. Malobabic, J.J. Liou, "Bidirectional devices for automotive-grade electrostatic discharge applications," *IEEE Electron Device Letters*, vol. 33, no. 6, pp. 860–862, 2012.

5. M. Stecher, N. Jensen, M. Denison, R. Rudolf, B. Strzalkoswi, M.N. Muenzer, L. Lorenz, "Key technologies for system-integration in the automotive and industrial applications," *IEEE Transactions on Power Electronics*, vol. 20, no. 3, pp. 537, 549, May 2005.

6. S. Malobabic, J.A. Salcedo, J.-J. Hajjar, J.J. Liou, "NLDMOS ESD scaling under human metal model for 40 V mixed-signal applications," *IEEE Electron Device Letters*, vol. 33, no. 11, pp. 1595, 1597, November 2012.

7. M.P.J. Mergens, M.T. Mayerhofer, J.A. Willemen, M. Stecher, "ESD protection considerations in advanced high-voltage technologies for automotive," in *Electrical Overstress/Electrostatic Discharge Symposium*, pp. 54–63, September 2006, https://www.esda.org/. [Accessed May 24, 2015.]

8. J.A. Salcedo, D. Clarke, J.-J. Hajjar, "On-chip protection for automotive integrated circuits robustness," in *IEEE International Caribbean Conference on Devices, Circuits and Systems (ICCDCS)*, pp. 1–5, March 2012, http://www.ieee.org/index.html. [Accessed May 24, 2015.]

9. J.A. Salcedo, H. Zhu, A.W. Righter, J.-J. Hajjar, "Electrostatic discharge protection framework for mixed-signal high voltage CMOS applications," in *IEEE International Conference on Solid-State and Integrated-Circuit Technology*, pp. 329–332, October 2008, http://www.ieee.org/index.html. [Accessed May 24, 2015.]

10. AEC-Q100: Automotive Electronics Counsel, *Device-Level ESD Immunity Standard*, 2003, http://www.aecouncil.com/AECDocuments.html.

11. IEC 61000-4-2, "Electromagnetic compatibility (EMC)—Part 4-2: Testing and measurement techniques—Electrostatic discharge immunity test," 2008.

12. ISO 10605, "Road vehicles—Test methods for electrical disturbances from electrostatic discharge," 2008.

13. ISO 7637-2, "Road vehicles—Electrical disturbances from conduction and coupling—Part 2: Electrical transient conduction along supply lines only," TC 22/SC 3, 2011, www.iso.org.

14. ISO 76371-1, "Road Vehicles Electrical Disturbances by conduction and coupling. Part 1: Passenger cars and light commercial vehicles with nominal 12 V supply voltage—Electrical transient conduction along supply lines only," 1990.

15. ISO 76371-2, "Road vehicles electrical disturbances by conduction and coupling. Part 2: Commercial vehicles with nominal 24 V supply voltage—Electrical transient conduction along supply lines only," 1990.

16. ISO 76371-3, "Road vehicles electrical disturbances by conduction and coupling. Part 3: Vehicles with nominal 12 V or 24 V supply voltage—Electrical transient transmission by capacitive or inductive coupling via lines other than the supply lines," 1995.

17. ISO 16750, "Road vehicles—Environmental conditions and testing for electrical and electronic equipment—Part 2: Electrical loads," 2010.

18. IEC 62132-4: International Electrotechnical Commission-IEC, "Integrated circuits—Measurement of electromagnetic immunity 150 kHz to 1 GHz—Part 4: Direct RF power injection (DPI) method," 2006.

19. ISO 11452-4: International Standards Organization, "Road vehicles—Component test methods for electrical disturbances from narrowband radiated electromagnetic energy—Part 4: Harness excitation methods," Bulk Current Injection (BCI) substitution method, 2011.

20. *LIN Network for Vehicle Applications*, SAE International, 2005, http://www.sae.org/. [Accessed May 24, 2015.]

21. Audi, BMW, Daimler, Porseche, Volkswagen, "Hardware requirements for LIN, CAN and FlexRay interfaces in automotive applications," 2011.

22. R. Rudolf et al., "Automotive 130 nm smart-power-technology including embedded flash functionality," in *IEEE International Symposium on Power Semiconductor Devices and ICs*, pp. 20–23, May 2011, http://www.ieee.org/index.html. [Accessed May 24, 2015.]

23. J.A. Salcedo, G. Cosgrave, D. Clarke, Y. Huang, "Protection system for integrated circuits and methods of forming the same," US Patent # 8,947,841, March 2015.

24. V.A. Vashchenko, W. Kindt, M. Ter Beek, P. Hopper, "Implementation of 60 V tolerant dual direction ESD protection in 5 V BiCMOS process for automotive application," in *Electrical Overstress/Electrostatic Discharge Symposium*, pp. 1–8, September 2004, https://www.esda.org/. [Accessed May 24, 2015.]

25. J.A. Salcedo, J.J. Liou, J.C. Bernier, "Design and integration of novel SCR-based devices for ESD protection in CMOS/BiCMOS technologies," *IEEE Transactions on Electron Devices*, vol. 52, no. 12, pp. 2682, 2689, December 2005.

26. M.-D. Ker, W.-J. Chang, "On-Chip ESD protection design for automotive vacuum-fluorescent-display (VFD) driver IC to sustain high ESD stress," *IEEE Transactions on Device and Materials Reliability*, vol. 7, no. 3, pp. 438, 445, September 2007.

27. J.A. Salcedo, J.J. Liou, Z. Liu, J.E. Vinson, "TCAD methodology for design of SCR devices for electrostatic discharge (ESD) applications," *IEEE Transactions on Electron Devices*, vol. 54, no. 4, pp. 822, 832, April 2007.

28. Y. Cao et al., "A failure levels study of non-snapback ESD devices for automotive applications," in *IEEE International Reliability Physics Symposium*, 2010, http://www.ieee.org/index.html. [Accessed May 24, 2015.]

29. P. Wessels et al., "Advanced 100 V 0.13 µm BCD process for next generation automotive applications," in *IEEE International Symposium on Power Semiconductor Devices & IC's*, 2006.

30. F. Magrini et al., "Advanced Wunsch-Bell based application for automotive pulse robustness sizing," in *Electrical Overstress/Electrostatic Discharge Symposium*, 2014, https://www.esda.org/. [Accessed May 24, 2015.]

31. A. Gendron et al., "New high voltage ESD protection devices based on bipolar transistors for automotive applications," in *EOS/ESD*, 2011, https://www.esda.org/. [Accessed May 24, 2015.]

32. A. Imbruglia et al., "BCD and discrete technologies for power management ICs development," in *ESTEC: European Space Components Conference*, 2013, http://www.esa.int/About_Us/ESTEC/European_Space_Research_and_Technology_Centre_ESTEC2. [Accessed May 24, 2015.]

33. J.-W. Lee et al., "Novel isolation ring structure for latch-up and power efficiency improvement of smart power IC's," in *Electrical Overstress/Electrostatic Discharge Symposium*, 2013, https://www.esda.org/. [Accessed May 24, 2015.]

34. R. Lerner et al., "A trench isolated thick SOI process as platform for various electrical and optical integrated devices," in *SOI-3D-Subthreshold Microelectronics Technology Unified Conference*, 2013, https://www.ieee.org/conferences_events/conferences/conferencedetails/index.html?Conf_ID=33072. [Accessed May 24, 2015.]

35. S. Malobabic, J.A. Salcedo, J.-J. Hajjar, J.J. Liou, "Analysis of safe operating area of NLDMOS and PLDMOS transistors subject to transient stresses," *IEEE Transactions on Electron Devices*, vol. 57, no. 10, pp. 2655–2663, October 2010.

36. J.A. Salcedo, "Design and characterization of novel devices for new generation of electrostatic discharge (ESD) protection structures," Doctoral Dissertation, 2006, http://etd.fcla.edu/CF/CFE0001213/Salcedo_Javier_A_200608_PhD.pdf.

37. S. Malobabic, J.A. Salcedo, A.W. Righter, J.-J. Hajjar, J.J. Liou, "A new ESD design methodology for high voltage DMOS applications," in *Electrical Overstress/Electrostatic Discharge Symposium*, ESD Association. pp. 1–10, October 2010, http://www.ieee.org/index.html. [Accessed May 24, 2015.]

38. D. Clarke, J.A. Salcedo, B. Moane, J. Luo, S. Murnane, K. Heffernan, J. Twomey, S. Heffernan, G. Cosgrave, "Junction-isolated blocking voltage devices with integrated protection structures and method of forming the same," US Patent 8,796,729, August 2014.

39. J.A. Salcedo, "Apparatus and method for electronics circuit protection," US Patent 8,553,380, October 2013.
40. J.A. Salcedo, D. H. Whitney, "Method for protecting electronic circuits operating under high stress conditions," US Patent 8,772,091, July 2014.
41. J.A. Salcedo, K. Sweetland, "Apparatus and method for transient electrical overstress protection," US Patent 8,466,489, June 2013, https://www.esda.org/. [Accessed May 24, 2015.]
42. *Analog Devices Automotive Audio Bus Technology*, http://www.analog.com.
43. J.A. Salcedo, "Low voltage protection devices for precision transceivers and methods of forming the same," US Patent 8,610,251, December 2013.
44. D.F. Ellis, Y. Zhou, J.A. Salcedo, J.-J. Hajjar, J.J. Liou, "Prediction and modeling of thin gate oxide breakdown subject to arbitrary transient stresses," *IEEE Transactions on Electron Devices*, vol. 57, no. 9, pp. 2296–2305, September 2010.
45. J.D.V. Iwaarden, A.V.D. Wide, A.R.T. Williams, B.G. Dale, "Quality management system development in the automotive sector," in *IEEE International Engineering Management Conference*, Vol. 3, pp. 1095, 1099, October 2004.
46. A. Ryan, "Quality systems in manufacturing," in *IEEE International Engineering Management Conference*, pp. 52, 55, August 2004, http://ieeexplore.ieee.org/search/searchresult.jsp?newsearch=true&queryText=Quality+systems+in+manufacturing. [Accessed May 24, 2015.]

7 ESD Sensitivity of GaN-Based Electronic Devices

Gaudenzio Meneghesso, Matteo Meneghini, and Enrico Zanoni

CONTENTS

This chapter reviews the main issues related to the electrostatic discharge (ESD) instabilities of electronic and optoelectronic devices based on gallium nitride; more specifically, we will describe the failure mechanisms of GaN-based transistors for radio frequency (RF) and power applications and of advanced light-emitting diodes (LEDs) for application in the general lighting field.

By summarizing the most relevant findings in this area, we show that GaN devices are intrinsically very robust to high electric fields and ESD; however, the presence of intrinsic weak points (such as extended defects and dislocations) and the use of a non-optimized layout may significantly reduce the failure threshold of GaN devices, thus leading to a premature failure. The typical results obtained on GaN-based transistors are critically compared with data obtained on other compound semiconductors (such as AlInGaP and GaAs) devices, to provide a more extensive description of the topic.

In this chapter, we also summarize the current status of the development of ESD protection structures based on GaN and the related technological problems.

7.1   INTRODUCTION ON GaN-BASED DEVICES

Over the last decade, GaN has demonstrated to be an excellent material for the fabrication of a variety of devices with outstanding performance, namely, (1) high-speed transistors for microwave applications; (2) high-voltage diodes and transistors for application in high-efficiency power conversion systems; and (3) LEDs and lasers emitting in the visible spectral range, which are finding wide application in the solid-state lighting field.

Gallium nitride is a wide band gap semiconductor, with an energy gap of 3.4 eV; thanks to this property, GaN-based devices can be operated at relatively high temperatures (>400°C) without becoming intrinsic; these temperature limits are considerably higher than those of silicon (150°C) and GaAs (<200°C). The possibility of reaching high temperatures allows one to increase the power dissipated on the individual devices (thus reducing the fabrication costs) or to reduce the size of the heat sinks (thus reducing system complexity, weight, and cost).

Another important property of GaN is the high saturation velocity (2.5×10^7 cm/s), compared to silicon (1×10^7 cm/s) and 4H-SiC (2×10^7 cm/s) [1]; this enables high-frequency operation, with direct benefits when the devices are used in the microwave field. Another advantage is that high-frequency operation permits to reduce the size of the passive components adopted in switching and RF systems.

GaN has also a high breakdown field (>3 MV/cm), compared to silicon (0.3 MV/cm) [2], thus permitting to fabricate high-voltage devices with a thinner layer of semiconductor. Thanks to this important property, GaN-based transistors and diodes with breakdown voltage in the 1–10 kV range have been already demonstrated, thus clearing the way for a massive penetration of GaN in the power electronics field.

Finally, GaN is a direct band gap semiconductor; this means that it has a high radiative efficiency and can be effectively used for the fabrication of high-performance LEDs and laser diodes (LDs) [3]. The emission wavelength can be tuned by using In-based or Al-based alloys (InGaN and AlGaN, respectively) [4]; devices with emission wavelengths ranging from the deep-ultraviolet to the red spectral region have already been demonstrated.

Ideally, GaN would be a perfect semiconductor in terms of ESD stability, thanks to its high breakdown field and maximum operating temperature [5,6]. However, issues related to material properties and device layout can significantly limit the ESD robustness of the devices [7–28]. One of the most relevant factors that limit the crystal quality of GaN is the relatively high density of threading dislocations (T_{dd}), which strongly depend on the substrate used for the growth of GaN. In fact, the lattice mismatch between GaN and the substrate can favor the generation of extended defects during the growth of the devices. Typical values of T_{dd} range between $10^5/cm^2$, in the case of growth on a native GaN substrate (this is usually the case of GaN-based LDs), and $10^7–10^9/cm^2$, in the case GaN is grown on a foreign substrate. The most commonly adopted foreign substrates are silicon carbide (SiC, usually adopted for RF transistors), sapphire (widely adopted for LEDs), and silicon (used for high-power transistors and LEDs).

Extended defects (such as dislocations) may act as conductive and localized current paths (i.e., weak spots) when the devices are submitted to ESD or EOS events [29,30]. Another relevant problem may arise from the fact that GaN-based devices may show relevant trapping effects [31]: the presence of charged defects may significantly influence the electric field within the devices, in a non-predictable way [15]. Finally, in some cases, the processing method for device fabrication may introduce further weak spots, related to imperfections in the mesa or contact shapes, and/or to the presence of sharp edges at the contacts [32].

In the following sections, we describe the most common failure mechanisms of GaN-based devices submitted to ESD events; the discussion is divided into two parts: the first one is on GaN-based transistors and the second one is on GaN optoelectronic devices.

7.2 GaN-BASED TRANSISTORS: DEVICE STRUCTURE AND ESD-RELATED ISSUES

Figure 7.1 shows the typical structure of a GaN-based high-electron mobility transistor (HEMT); the structure is grown on a foreign substrate and consists of a GaN channel layer and an AlGaN barrier layer, on which a Schottky (or MIS) contact is fabricated, to control the flow of charge in the channel. Source and drain ohmic contacts and a dielectric passivation are then introduced to finalize device fabrication. GaN is a polar material: as a consequence, the use of an AlGaN/GaN heterostructure results in the generation of a bidimensional electron gas (2DEG) with extremely high mobility. Most of the papers investigate the ESD robustness of GaN HEMTs by means of transmission line pulse (TLP) and human body model (HBM) measurements.

7.2.1 Hard and Soft Failure in GaN HEMTs

Typical results of a TLP test carried out on a GaN HEMT are shown in Figure 7.2 [14]; here the TLP stress was applied to the drain, with source grounded and gate floating. This is one of the most critical conditions and represents one of the typical ESD events that may occur during device manufacturing and handling. During the TLP test, this device—which was grown on a SiC substrate—shows the typical transistor I_D-V_D curve, until it reaches the hard failure (in this case at 85 V, 1.2 A), which

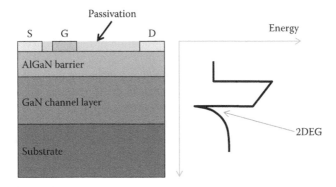

FIGURE 7.1 (Left) Typical structure of a GaN-based transistor; (Right) diagram of the conduction band of the same device, indicating the presence of a bidimensional electron gas (2DEG).

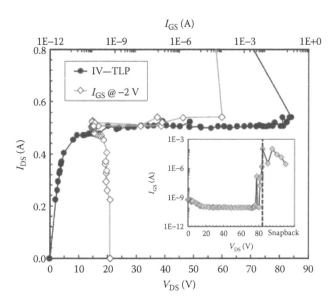

FIGURE 7.2   TLP characterization carried out on a GaN-based HEMT with $W = 1$ mm. (Data from Tazzoli, A. et al., "ESD robustness of AlGaN/GaN HEMT devices," in *Proceedings of the EOS/ESD Symposium*, 2007, 4A.4-1.) TLP test was applied between drain and source, with gate floating.

leads to a sudden decrease in the equivalent impedance. A closer look to device characteristics (Figure 7.3) reveals that—close to the failure limit—TLP testing may induce changes in the subthreshold current of the devices, which can be ascribed to the degradation of the gate Schottky diode.

The results indicate that TLP testing applied between drain and source may induce a severe degradation of the gate junction, even when the gate terminal is kept floating during the stress. This result was interpreted by considering the capacitive coupling between the drain, gate, and source terminals during the TLP events. It was demonstrated that, because of the fast rise time of the TLPs (below 1 ns), the floating gate can follow the drain pulse (Figure 7.4), thus reaching high-voltage levels and leading to a degradation of the I–V curves of the gate diode [14].

Besides gate leakage current, other parameters—including threshold voltage and open channel resistance—can show a parametric (soft) degradation when GaN-based transistors are submitted to ESD pulses with increasing amplitude. An example is reported in Figure 7.5, which shows the changes in the threshold voltage, open channel resistance, and source + drain resistances induced by the exposure to TLP tests with increasing current amplitude [8]. The TLP tests were carried out with floating (Figure 7.5a) and grounded (Figure 7.5b) gate. It can be noticed that—when the gate is kept floating during stress—TLP testing can induce a change in the threshold voltage (in the particular case reported in Figure 7.5a, V_T varies from −0.9 to −0.4 V); this change occurs during the initial stages of the TLP test and has been ascribed to

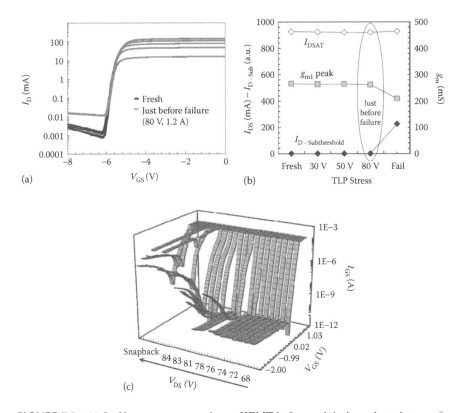

FIGURE 7.3 (a) I_D–V_G curves measured on a HEMT before and during selected steps of a TLP experiment; (b) variation of representative device parameters (saturation current I_{Dsat}, subthreshold current $I_{D-subthreshold}$, and transconductance peak, $g_{m1-peak}$) during the same stress experiment; (c) changes in the gate-source Schottky diode during the TLP stress. (Data from Tazzoli, A. et al., "ESD robustness of AlGaN/GaN HEMT devices," in *Proceedings of the EOS/ESD Symposium*, 2007, 4A.4-1.)

the trapping of electrons in the buffer side of the channel-buffer interface. On the other hand, the changes in open channel resistance (R_O, Figure 7.5a; $I_{stress} > 1$ A) have been ascribed to a gradual degradation of the low-field electron mobility.

7.2.2 INVESTIGATION OF THE PHYSICAL ORIGIN OF FAILURE

Several techniques have been used to investigate the physical origin of the failure of GaN-based HEMTs submitted to ESD events. Kuzmik et al. [8] presented an extensive investigation of the failure of GaN-based HEMTs submitted to ESD; the results of backside infrared camera measurements (Figure 7.6) indicate that—as a consequence of TLP testing—dark spots are generated on the grounded contact. This result has been interpreted by considering that not only temperature but also current flow direction is decisive for the formation of physically damaged regions [8].

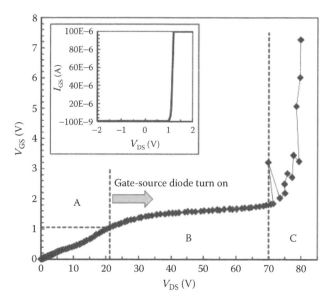

FIGURE 7.4 Change in the gate voltage induced by the increase in the TLP voltage applied to the drain of the device (TLP stress is applied between drain and source, with gate floating). As can be noticed, the gate-source bias can significantly increase because of the capacitive coupling between gate and drain terminals. (Data from Tazzoli, A. et al., "ESD robustness of AlGaN/GaN HEMT devices," in *Proceedings of the EOS/ESD Symposium*, 2007, 4A.4-1.)

According to [8] and [33], this is an indication that a high flow of electrons perpendicular to the surface generates an electromigration effect and the consequent formation of dark spots.

Electroluminescence (EL) represents another powerful tool for the analysis of the leakage paths generated after TLP testing on GaN-based HEMTs; as discussed earlier, even nondestructive pulses may lead to a significant increase in the leakage current of the gate Schottky diode. In EL measurements, this corresponds to the generation of hot spots, that is, of localized shunt paths, which directly connect the gate metal to the channel layer (see an example in Figure 7.7). Possible mechanisms responsible for the generation of hot spots are (1) the presence of defects (such as V-shaped defects) related to threading dislocations, which locally reduce the thickness (and the breakdown voltage) of the AlGaN barrier layer, thus leading to the formation of localized weak region (a representative example is shown in Figure 7.8a). When submitted to high-current/voltage TLPs, these defective regions may rapidly fail, thus generating a nanometer-size conductive path between gate and channel. (2) Another process that can lead to a localized failure when the devices are submitted to high voltages is the converse piezoelectric effect [34]. When a high drain bias is applied between gate and drain, a significant tensile strain builds up in proximity of the gate edge (on the drain side), which

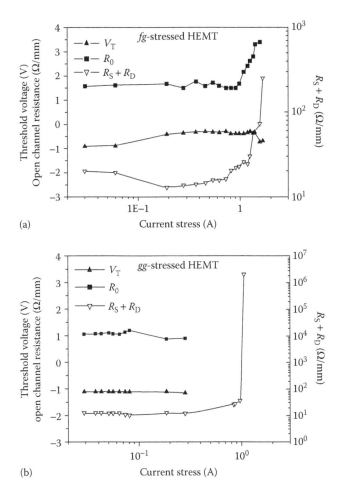

FIGURE 7.5 Changes in the threshold voltage of AlGaN/GaN HEMT, open channel resistance, and source + drain resistances values in dependence of the ESD stress current level; gate is (a) floating during the stress and (b) grounded during the stress. (Reprinted from *Solid-State Electronics*, 48, Kuzmík, J. et al., Electrical overstress in AlGaN/GaN HEMTs: Study of degradation processes, 271–276, Copyright 2004, with permission from Elsevier.)

is the region where the electric field is maximum (see Figure 7.8b). When the applied bias exceeds a "critical voltage," the excess elastic energy accumulated by the device is released through the formation of localized crystallographic defects, whose position can be detected by spatially resolved EL measurements [35] and which act as localized shunt paths [36].

A detailed description of the dynamics of electric field distribution during ESD tests can be obtained based on transient interferometric mapping (TIM) measurements; this technique provides information on single-event dynamics, without

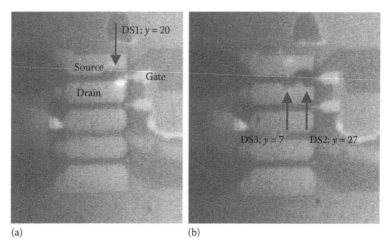

(a)   (b)

FIGURE 7.6 Backside infrared camera view of the AlGaN/GaN HEMT after $I_{stress} = 0.25$ A on the (a) drain and (b) source. Arrows mark dark spot appearance. Only labeled HEMT electrodes were connected. (Reprinted from *Solid-State Electronics*, 48, Kuzmík, J. et al., Electrical overstress in AlGaN/GaN HEMTs: Study of degradation processes, 271–276, Copyright 2004, with permission from Elsevier.)

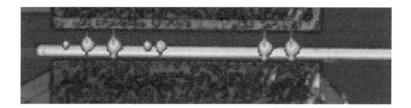

FIGURE 7.7 Spatially resolved electroluminescence measurements carried out on a GaN-based HEMT after 0.2 A nondestructive TLP test. The hot spots represent preferential paths for leakage current conduction generated by the high voltage (>150 V) applied to the devices.

the need of carrying out repetitive stressing [15]; by carrying out TIM measurements, Bychikhin et al. [15] demonstrated that—for high voltages—the electric field becomes inhomogeneous, because of the trapping of electrons at surface traps (Figure 7.9a); they also demonstrated that the capture of electrons is a fast process, with time constants smaller than 100 ns (i.e., comparable to the duration of ESD events) (Figure 7.9b). Based on TLP measurements and finite element thermal simulations, they indicated that the onset of breakdown is not activated by self-heating but, more likely, by the injection of a critical amount of electrons into trap states and by the consequent increase in the electric field. Figure 7.9 reports typical results of TIM measurements carried out on TLM structures; the results provide a description of the spatial distribution of the electric field as a function of time, after a high bias has been applied to the devices. From these

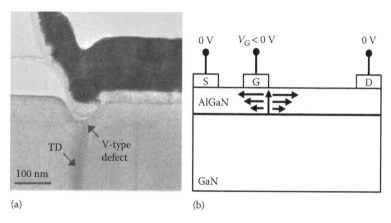

(a)   (b)

FIGURE 7.8 (a) TEM of a defective region with a V-defect next to the gate of a transistor. A threading dislocation with the corresponding defect is clearly visible. (Courtesy of David Cullen and David Smith, Arizona State University, Phoenix, Arizona; Data from Cullen, D. et al., *IEEE Trans. Device Mater. Reliab.*, 13, 126, 2013.) (b) A schematic representation of the failure process related to converse piezoelectric effect. (Data from Joh, J. and del Alamo, J.A., "Mechanisms for electrical degradation of GaN high-electron mobility transistors," in *IEEE IEDM Technical Digest*, 2006.)

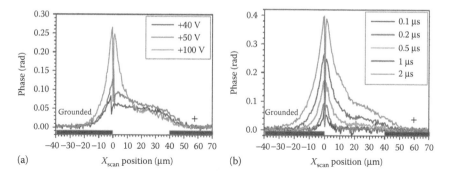

(a)   (b)

FIGURE 7.9 Results of TIM measurements reflecting the distribution of the electric field as a function of (a) cathode–anode voltage at $t = 1$ μs and (b) time at $V = 100$ V; $L = 40$ μm. The electrons are injected from the electrode on the left; they cause the depletion of the channel and enhance the electric field. (Data from Bychikhin, S. et al., "Investigation of nanosecond-time-scale dynamics of electric field distribution and breakdown phenomena in InAlN/GaN TLM structures," in *Proceedings of the European Workshop on Heterostructure Technology*, 2009; Courtesy of Dyoniz Pogany, University of Wien, Vienna, Austria.)

results, it is clear that trapping processes may significantly affect the distribution and the amplitude of the electric field; the engineering of the electric field (via suitable field plates and passivation [37]) and the control of trapping processes represent important steps toward the fabrication of reliable and robust gallium nitride transistors.

7.2.3 LATERAL VERSUS VERTICAL ESD EFFECTS

Recent reports [13] demonstrated that ESD events may induce two different kinds of failure processes, depending on the layout and geometry of the devices: lateral (gate-to-drain) failure and vertical (drain-to-substrate) failure. An extensive description of these two mechanisms was recently presented by Chen et al. in [13], based on the investigation of the HBM and TLP I–V characteristics of Schottky diodes based on GaN-on-Si technology. The diodes were fabricated by metal-organic chemical vapor deposition (MOCVD), starting from a silicon substrate. Device structure consisted of a 2 μm AlGaN buffer layer, a 150 nm GaN channel layer, and a 7 nm AlGaN channel layer. A Ni/Au Schottky contact and a Ti/Al/Mo/Au ohmic contact were then fabricated; to investigate the impact of layout and geometry on the ESD robustness of the devices, devices with varying anode-to-cathode lengths (L, between 1.5 and 20 μm) were investigated.

Lateral failure of the devices was investigated by carrying out forward-mode ESD investigation; the analysis was carried out with an on-wafer HBM tester and indicated that devices with short anode–cathode length (e.g., 1.5 μm) show an abrupt increase in current (corresponding to device failure), once a 2.4 kV voltage is reached (see Figure 7.10a). On the contrary, devices with long anode–cathode spacing show (for HBM voltages higher than 2 kV) a gradual (soft) increase in the leakage current (see Figure 7.10b), followed (for a HBM voltage of 3 kV) by a hard failure. A similar behavior was confirmed also by TLP tests. The different behavior of short- and long-spacing Schottky diodes can be explained as follows: in short-length diodes, the high electric field and high current density induce a high local thermal energy, which leads to a direct catastrophic damage of the Schottky contact.

On the other hand, TIM measurements were carried out to understand the different behavior of devices with long anode–cathode lengths. The phase shift of TIM measurements represents the local thermal energy density and the density of heat dissipation [13]. The results (Figure 7.11) indicated that short diodes have a nonuniform phase distribution, which originates from current crowding, which was indicated as responsible for hard failure. On the other hand, longer (15 μm) diodes have a more uniform distribution, indicating a substantial absence of crowding. The soft failure of long-length diodes was then tentatively ascribed to the degradation of the Schottky contact.

Reverse-bias ESD testing on diodes with variable anode–cathode spacing allowed Chen et al. [13] to separately investigate lateral failure from vertical breakdown mechanisms. Representative results of this investigation are shown in Figure 7.12, which reports the dependence of (reverse-mode) HBM ESD robustness of Schottky diodes with varying anode-to-cathode length. For short devices (<5 μm), the robustness of the diodes increases linearly with anode–cathode spacing. In this case, the breakdown voltage and the ESD robustness of the diodes is determined by the lateral field between anode and cathode. On the other hand, for longer lengths (>5 μm), the HBM robustness saturates around 350 V. This result suggests that for long devices the ESD robustness is dominated by the vertical electric field and that breakdown occurs vertically, that is, in the silicon substrate [38]. A significantly improved behavior can be obtained by removing the silicon substrate [38,39], as shown in Figure 7.12.

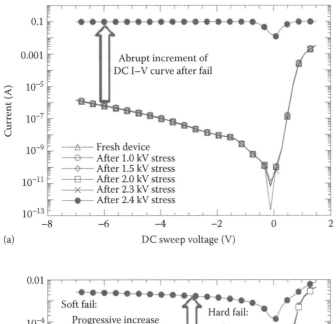

(a)

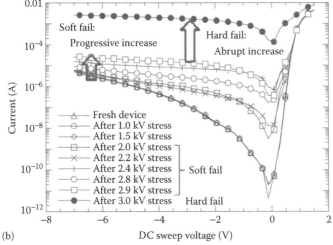

(b)

FIGURE 7.10 DC I–V characteristics (voltages sweep on normal operating region) after each HBM pre-charge voltage for Schottky diodes with 800 μm width and length of (a) 1.5 μm and (b) 15 μm. (Reproduced with permission from Chen, S.H. et al., *IEEE Trans. Device Mater. Reliab.*, 12, 589, 2012. Copyright 2012 IEEE.)

7.2.4 GaN-Based ESD Protection Structures

The development of protection structures for GaN devices is particularly difficult because of a set of intrinsic limitations: the first important issue is the fact that GaN technology—contrary to silicon—is not fully standardized, and for this reason it is difficult to develop standard protection devices. The processing of GaN-based

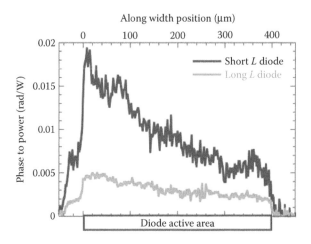

FIGURE 7.11 Normalized phase distributions of diode with 15 μm (long) and 1.5 μm (short) anode-to-cathode length and 800 μm width under 1. A 100 ns TLP stress. (Reproduced with permission from Chen, S.H. et al., *IEEE Trans. Device Mater. Reliab.*, 12, 589, 2012. Copyright 2012 IEEE.)

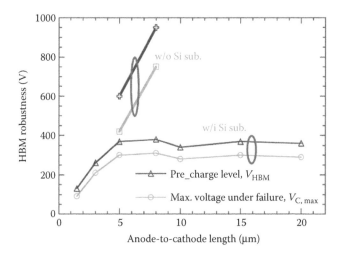

FIGURE 7.12 HBM ESD robustness and $V_{C,max}$ as a function of the length of Schottky diodes (reverse mode) on Si substrate with 800 μm width. (Reproduced with permission from Chen, S.H. et al., *IEEE Trans. Device Mater. Reliab.*, 12, 589, 2012. Copyright 2012 IEEE.)

monolithic microwave integrated circuits (MMICs)—usually adopting on a gold-based technology—may lead to the generation of metal spikes and poor edge definition; this may have a negative impact on the ESD robustness and lead to a premature failure of the devices. Other problems may arise from the fact that power transistors based on GaN are mainly grown on silicon, which has a poor lattice matching with

GaN; this may lead to high densities of structural defects and lower the threshold for the catastrophic failure of the devices.

The design of ESD protection structures is limited by the difficulty of using p-type doping in GaN-based HEMT processing; moreover, the low mobility of holes can limit the maximum speed and current levels reached by bipolar protection structures (such as silicon-controlled rectifiers [SCRs]).

The electronic devices that can be used as protection structures for GaN-based devices and circuits are (1) GaN-based transistors, (2) GaN-based Schottky diodes, and (3) resistors. It is worth noticing that as GaN-based devices have short response time and high switching performance, high-speed protection devices are required.

One of the most recent reports on GaN-based ESD protection devices has been presented by Wang et al. in [12]. In this paper, they proposed and fabricated a robust gallium nitride-based ESD protection structure, which consists of a depletion-mode pHEMT, a trigger diode chain, a pinch-off diode chain, and a current limiting resistor. A schematic structure of this device is shown in Figure 7.13: The GaN-based pHEMT is grown on a silicon substrate; the device is fabricated in a 0.35 μm technology. The main applications are the ESD protection of low-noise and power amplifiers for wireless communications, with operating voltage ranging between 2 and 4 V [40]. The gate of the transistor is connected to the anode of the protection structure through a trigger diode chain, which turns on the pHEMT when the pin is submitted to an ESD event. On the other hand, the pinch-off diode chain has the function of keeping the pHEMT in the off-state during normal circuit operation. In addition, it limits the leakage current flowing through the pHEMT during normal operation. During an ESD event, the voltage at the anode becomes higher than the turn-on voltage of the trigger diode chain; as a consequence, current starts to flow through the current limiting resistor, thus turning on the transistor. As a consequence, the ESD-induced current is shunted by the pHEMT toward the cathode contact [12]. This structure was fully characterized in terms of trigger voltage, leakage current, on-resistance (R_{on}), and TLP failure current (I_{t2}): Typical results obtained by using four diodes in the trigger diode chain and four diodes in the pinch-off diode chain are trigger voltage = 6 V, leakage current = 1.93×10^{-6} A, $R_{on} = 3.36 - 335$ Ω, and $I_{t2} = 3$ A.

This protection structure was submitted to detailed failure analysis, after TLP investigation: The results (Figure 7.14) demonstrated that the exposure to ESD events induced damages on the pinch-off diode chain; a more detailed investigation revealed that failures were located close to the anode sides of the pinch-off diodes, indicating that current crowding occurs in these regions. On the other hand, no damage was found on the pHEMT and trigger diode chain after ESD testing.

A similar protection device has been proposed also by Chiu et al. [17], for electric vehicle applications. This device consists of a five-diode chain, a GaN D-mode HEMT, and a current limiter resistor (in the kΩ range). The HEMT device has a dual-gate structure; this improves the ESD protection performance compared to single-gate devices. The trigger voltage of this structure is 510 V, while the failure current is 8.93 A.

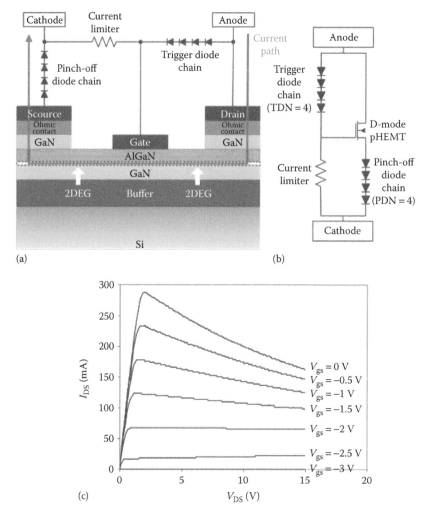

FIGURE 7.13   (a) Cross-sectional view of the proposed ESD protection clamp, consisting of a single-gate D-mode GaN pHEMT, a trigger diode chain, a pinch-off diode chain, and a resistor (current limiter). Solid line: the current path after the clamp is turned on. (b) Equivalent circuit of the GaN pHEMT–based ESD clamp. (c) Measured steady-state I–V curves of the pHEMT. (Reproduced with permission from Wang, Z. et al., *IEEE Electron Device Lett.*, 34, 1491, 2013. Copyright 2013 IEEE.)

7.2.5 ESD Stability of GaAs-Based Devices

Even if this chapter is focused on GaN-based devices, it is important to present a brief summary of the ESD issues of devices based on another important compound semiconductor, namely, GaAs. The electrical, optical, and thermal properties of GaAs are significantly different from those of GaN and silicon; more specifically:

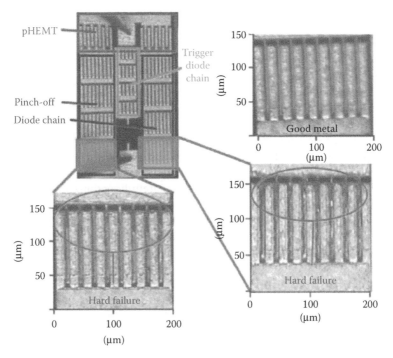

FIGURE 7.14 Photograph of the GaN pHEMT clamp, with enlarged pictures showing hard failures on the pinch-off diode chain. (Reproduced with permission from Wang, Z. et al., *IEEE Electron Device Lett.*, 34, 1491, 2013. Copyright 2013 IEEE.)

1. GaAs has a lower melting point (1238°C), compared to silicon (1414°C) and gallium nitride (>2500°C).
2. The thermal conductivity of GaAs (0.46 W/cm K) is lower than that of silicon (1.5 W/cm K) and gallium nitride (1.7 W/cm K) [41].
3. The breakdown field of GaAs (3.5 × 10⁵ V/cm) is comparable to that of silicon (2.5 × 10⁵ V/cm) and significantly lower than that of GaN (3.3 × 10⁵ V/cm).

This means that, compared to Si- and GaN-based transistors, GaAs devices heat up faster and melt earlier; moreover, compared to GaN-based devices, GaAs transistors fail at a lower voltage (for the same geometry), because of the differences in the breakdown field.

Several degradation mechanisms of GaAs HEMTs and MESFET have been presented so far in the literature. As described in [42], one of the most critical aspects is the fact that planar devices require the use of several metal layers, for the fabrication of the ohmic and Schottky contacts. The metal lines can show a lifetime limitation, because of the migration of the metal through the surface of the semiconductor. These mechanisms, which may lead to the failure of the devices, are significantly accelerated at high temperatures, electric field, and current densities, that is, in the conditions that take place during an ESD event. Common degradation mechanisms due to the interaction of metals (such as Au or Ni) with GaAs include (1) the lateral migration of the metal across the GaAs surface, which eventually leads to a short circuit [42–44]; (2) contact spiking, caused by focused current flow [42,44]; and (3) charge injection and oxide (dielectric) breakdown [42,44].

Another possible failure mechanisms is the filamentation in the semiconductor material: a localized current flow creates a melt filament; after the event, the damage region re-crystallizes, creating a resistive shunt path in parallel to the device. As described in [42], electromigration effects play a strong role also in other compound semiconductors such as InP and GaSb; since mass transport is easier—especially at low temperatures—if lattice defects (dislocations, interfaces, grain boundaries) are present [42], the quality of the crystal must be carefully optimized to ensure high robustness to the devices.

The lateral migration of metal has been observed in GaAs-based FET devices (MESFETs and HEMTs): During an ESD event, the commonly adopted gold metallization can interdiffuse with the GaAs substrate [42].

Another important issue of GaAs-based devices is the degradation of the Schottky contacts, which—in some cases—can lead to the blow off (explosion) of the gate metallization [42,45,46]. The optimization in gate layout and the adoption of suitable diffusion barriers allow to improve the reliability of the Schottky contacts.

On the other hand, the most critical aspect of heterojunction bipolar transistors (HBTs) is semiconductor damage by filamentation [42,47]. As the base of a HBT is very thin, several current paths are possible during an ESD event; the most probable is the conduction through the emitter–base junction, which leads to the failure of the device. In HBTs, the density of defects in the semiconductor is responsible for the burnout of the base layer [42].

7.2.6   ESD FAILURE OF GaN-BASED LEDs

Over the last few years, GaN-based LEDs have demonstrated to be excellent devices for the fabrication of high efficiency lamps; the direct band gap of GaN results in a high radiative efficiency, while by changing the indium content in the quantum wells (QWs), it is possible to tune the emitted wavelength in the visible and UV spectral range. In this section, we summarize the main ESD issues of GaN-based LEDs and present a comparison with LEDs based on other compound semiconductors (such as AlInGaP).

GaN-based LEDs have a quite complex structure: light emission occurs within InGaN (AlGaN in the case of UV LEDs) QWs, which have a typical thickness of few nanometers. The carriers are injected into the QWs from GaN barrier layers; the electrons and holes are injected from a Si-doped (n-type) region and a Mg-doped (p-type) region, when a positive bias is applied to the devices. As in the case of HEMTs, GaN-based LEDs are grown on a foreign substrate, because of the lack of native GaN wafers of reasonable size and cost. The most commonly adopted substrate materials are sapphire, silicon carbide, and silicon. The use of hetero-epitaxial growth—which may result in high density of structural defects, the small thickness

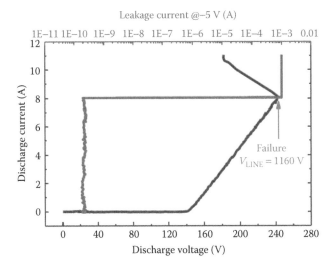

Leakage current @–5 V (A)

FIGURE 7.15 An example of the TLP measurement result on a GaN-based LED. The black line represents the pulsed I–V characteristic, while the gray line shows the trend of the leakage current at –5 V during the test. (Reprinted from *Microelectronics Reliability*, 54, Dal Lago, M. et al., ESD on GaN-based LEDs: An analysis based on dynamic electroluminescence measurements and current waveforms, 2138–2141, Copyright 2014, with permission from Elsevier.)

of the QW, and the high electric fields within the QW region (in the order of MV/cm) may significantly limit the ESD robustness of GaN-based LEDs.

Figure 7.15 reports the typical results of a TLP test carried out—in reverse bias—on a GaN-based blue LED; at high-voltage levels ($V_{dut} > 140$ V), the diode junction starts showing a measurable (reverse) breakdown current, with a high series resistance. Failure current is approximately 8 A, while failure voltage is 240 V. The same figure reports also the leakage current measured after each ESD event; failure consists of a sudden increase in leakage, which leads to the shortening of the junction.

One of the earliest reports on the effects of ESD on the electrical performance of InGaN-based LEDs was presented in 2001 by Meneghesso et al. [48]. They reported on the results of ESD testing (both HBM and TLP) of commercial GaN LEDs, fabricated by three different suppliers. They investigated LEDs grown both on sapphire and on SiC substrate. The difference between these two sets of samples in the direction of current flow is as SiC is a conductive substrate, devices grown on silicon carbide have a vertical current flow. The current is injected from the anode (placed on the top of the device) and collected from the cathode (located at the bottom), through the conductive SiC substrate. On the other hand, devices grown on a sapphire substrate have a horizontal (lateral) layout; as sapphire is insulating, current cannot be collected through the substrate, and the cathode terminal is placed on the surface of the device (suitable mesas are fabricated to provide electrical isolation between the anode and the cathode contacts).

FIGURE 7.16 Optical micrograph of an LED failed after reverse HBM ESD testing. (Data from Meneghesso, G. et al., "Electrostatic discharge and electrical overstress on GaN/InGaN light emitting diodes," in *Proceedings of the EOS/ESD Symposium*, September 11–13, 2001, pp. 247, 252.)

In [48], Meneghesso et al. demonstrated that the failure threshold of commercial devices can show a strong variability, ranging from 1 kV HBM (~0.5 A TLP) to more than 8 kV HBM (5 A TLP). The devices with high initial (reverse) leakage current were found to have a low HBM failure threshold. Optical (Figure 7.16) and electron (Figure 7.17) microscopy were carried out to identify the position of the regions affected by ESD damage; the results demonstrated that on these "weak" samples, the failure sites are distributed randomly on device area. This fact suggested that failure occurs in proximity of weak spots, related to the presence of defects that can promote junction short circuiting and metal interdiffusion.

The results in [48] also demonstrated that—in LEDs grown on a sapphire substrate—a poor layout may result in a significant current crowding (which can be identified by means of emission microscopy); this can lead to a premature failure of the devices, because of the fusion of the top-side semitransparent contact. This process is more prominent at the edges of the emitting area. On the other hand, the devices grown on silicon carbide substrate (which have a vertical current flow) showed a better robustness to ESD (up to 8 kV HBM and 6 A TLP).

The ESD robustness of LEDs depends strongly on the semiconductor material used for device fabrication. Figures 7.18 and 7.19 compare the ESD robustness of

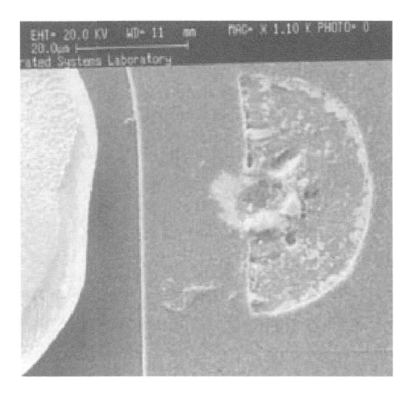

FIGURE 7.17 Scanning electron microscopy image of an LED failed after reverse-bias TLP test. (Data from Meneghesso, G. et al., "Electrostatic discharge and electrical overstress on GaN/InGaN light emitting diodes," in *Proceedings of the EOS/ESD Symposium*, September 11–13, 2001, pp. 247, 252.)

GaN-based blue and green LEDs with that of conventional red LEDs based on AlInGaP. The tests have been carried out on for different sets of devices (fabricated by different manufacturers), both under forward and reverse ESD conditions. Figures 7.18 and 7.19 show that InGaN-based (green and blue) LEDs have a lower robustness with respect to AlInGaP (red) devices. This result indicates that the semiconductor material used for the fabrication of LEDs may significantly impact on the robustness of the devices. In addition, the LEDs are in general more sensitive to reverse-bias ESD testing, with respect to forward-bias stress (see also Figure 7.20); this is due to the fact that—during a reverse-bias ESD event—current flows only through localized paths, which are related to the presence of extended defects. In proximity of these (nanometer-sized) weak spots, current density may become very high, thus leading to the catastrophic failure of the devices. On the contrary, during a forward-bias ESD event, current can spread on the whole device area, because the p–n junction is forward biased. This significantly reduces the (local) current density and results in a stronger ESD robustness under forward bias.

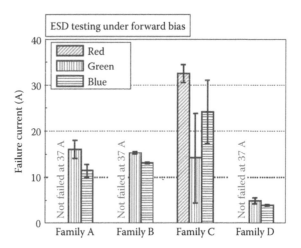

FIGURE 7.18 TLP failure current of four different sets of red, green, and blue LEDs. Results are referred to forward-bias TLP tests. Failure criterion is an increase in leakage current (measured at 1 V) above 10 µA. (Reprinted from *Microelectronics Reliability*, 54, Meneghini, M. et al., ESD degradation and robustness of RGB LEDs and modules: An investigation based on combined electrical and optical measurements, 1143, Copyright 2014, with permission from Elsevier.)

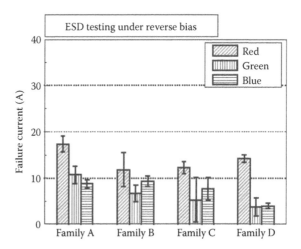

FIGURE 7.19 TLP failure current of four different sets of red, green, and blue LEDs. Results are referred to reverse-bias TLP tests. Failure criterion is an increase of leakage current (measured at −5 V) above 10 µA. (Reprinted from *Microelectronics Reliability*, 54, Meneghini, M. et al., ESD degradation and robustness of RGB LEDs and modules: An investigation based on combined electrical and optical measurements, 1143, Copyright 2014, with permission from Elsevier.)

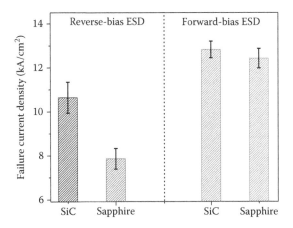

FIGURE 7.20 TLP failure current of LEDs grown on two different substrates, SiC and sapphire. Results obtained with both reverse- and forward-bias ESD test are reported. (Reproduced with permission from Meneghini, M. et al., *IEEE Trans. Electron Devices*, 57, 108–118, 2010. Copyright 2010 IEEE.)

Due to the important role of extended defects in determining the failure of the devices, in most of the cases the (reverse-bias) ESD robustness of LEDs depends on the quality of the material; the choice of the substrate material (SiC, sapphire, etc.) can significantly impact on the density of extended defects, which may act as preferential paths for leakage current conduction. A higher lattice mismatch may—in principle—result in a worst ESD instability; Figure 7.20 shows an example of a study carried out on LEDs grown on different substrates (SiC and sapphire). Under reverse-bias conditions, the LEDs grown on sapphire (higher lattice mismatch with GaN) have an ESD robustness lower than LEDs grown on a SiC substrate (lower lattice mismatch). This difference is not present under forward-bias ESD, because under positive bias, current is not focused on defect sites but spreads on the whole device area.

7.2.7 SOFT AND HARD FAILURE OF GaN-BASED LEDs

In general, ESD events may result in the catastrophic failure (i.e., shortening) of the devices. However, recent studies [18,21] demonstrated that nondestructive ESD events may also lead to a soft failure, that is, to changes in the electrical and optical characteristics of the devices. A first example is reported in Figure 7.21, which shows that nondestructive ESD pulses with increasing amplitude may induce a decrease in the leakage current of the devices (see especially the curves related to the green sample). This behavior has been explained by considering that during a reverse-bias ESD event, an extremely high-current density flows through a small number of preferential leakage paths, which are related to the presence of structural defects [30], and whose position can be identified by means of EL microscopy. During ESD events with extremely high current, some of these leakage paths may be annihilated (due to the high energy dissipated during the pulse), and this may result

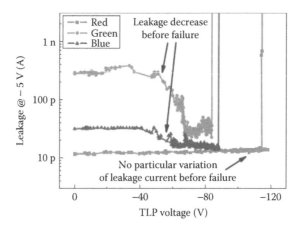

FIGURE 7.21   Changes in the leakage current in three different LEDs submitted to ESD testing. This figure clearly shows the decrease of the leakage current in GaN-based green and blue LEDs for negative pulses smaller than the failure threshold. (Reprinted from *Microelectronics Reliability*, 53, Vaccari, S. et al., ESD characterization of multi-chip RGB LEDs, 1510–1513, Copyright 2013, with permission from Elsevier.)

in a significant decrease in the (reverse-bias) leakage current (see [19,21] for details). A clear description of this mechanism is given in Figure 7.22, which shows the false-color EL pattern measured under reverse-bias conditions on a green LED. The luminescent spots indicate the location of the leakage paths. As shown in Figure 7.22b and c, some of the leakage paths may be annihilated after stress, thus leading to a change in the leakage current of the devices.

This is not the only mechanism that can lead to a "soft failure" of LEDs submitted to ESD testing; another mechanism consists of the generation of parasitic shunt paths in parallel to the junction; an example is shown in Figure 7.23, which reports the optical power versus current curves measured on an LED after applying ESD pulses with increasing amplitude. As can be noticed, the exposure to nondestructive ESD events leads to a partial decrease in the optical power emitted by the devices. This mechanism does not induce a total quenching (i.e., catastrophic failure) of the devices, but just a gradual change in the optical output. It is caused by the decrease in the (shunt) parallel resistance of the LED, because of the generation of defects within the semiconductor material.

7.2.8   Improving the ESD Robustness of GaN-Based LEDs

Over the last few years, several methods [52,53] for improving the ESD robustness of GaN-based LEDs have been proposed. In this section, we review the most interesting or widely adopted solutions. A first possibility is to add—in parallel to each LED—an external protection device, such as a diode or a transient voltage suppressor; these

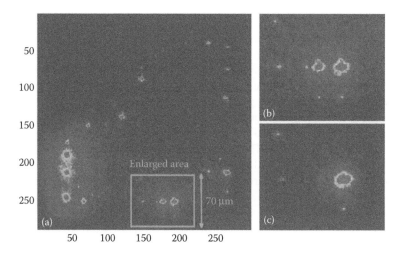

FIGURE 7.22 (a) False-color emission image showing the reverse-bias EL pattern of one green LED before the ESD test. (b) Enlarged image of the area highlighted in (a); (c) enlarged image of the same area of (b), taken after nondestructive TLP test (same integration time and false-color scale are used in both images). (Reprinted from *Microelectronics Reliability*, 53, Vaccari, S. et al., ESD characterization of multi-chip RGB LEDs, 1510–1513, Copyright 2013, with permission from Elsevier.)

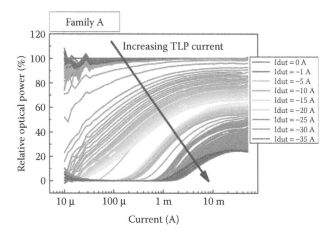

FIGURE 7.23 Changes in the optical power versus current curves of an LED measured at different stages of a TLP test. Curves are normalized with respect to the optical power before ESD test. (Reprinted from *Microelectronics Reliability*, 53, Vaccari, S. et al., ESD characterization of multi-chip RGB LEDs, 1510–1513, Copyright 2013, with permission from Elsevier.)

are usually silicon devices, which can be either unipolar or bipolar. In most of the cases, these devices are integrated in the LED package, to ensure small-size and optimal integration. Another possibility is to use metal-oxide varistors (MOVs), which are voltage-dependent resistors used to protect electronic devices from very short–voltage spikes. MOVs are usually placed at the input of a driving circuit, but, in some cases, they can also be placed on the output side [49].

The use of external protection devices may increase the complexity of the system and the related cost. For this reason, several researchers have recently proposed possible methods to improve the ESD robustness of the LED chips, as an alternative to the adoption of external ESD protection devices. A first solution was proposed by Chuang et al. [24], who fabricated a GaN-based LED incorporating an MOS capacitor. The LED/MOS device was submitted to HBM testing to prove the efficacy of this solution. The results demonstrated that when the MOS capacitor is connected in parallel to the LED, the ESD robustness is increased from 200 to 1500 V HBM. Moreover, this solution does not imply any significant variation in the light extraction efficiency.

Another approach that results in an improvement of the ESD stability of GaN-based LEDs is the use of an internal GaN Schottky diode, placed inside the LED chip. Chang et al. [50] fabricated internal GaN Schottky diodes after the epitaxial growth of GaN lateral LEDs (grown on a sapphire substrate). They adopted an etching and re-deposition technique; by properly selecting the etching regions (located below the bonding pads), it is possible to minimize the optical losses related to the etching procedure. This strategy allowed to increase the ESD failure threshold of the LEDs from 450 to 1300 V but resulted in increased reverse current and forward turn-on voltage. More recently, Jeong et al. [26] proposed a method for integrating an internal protection diode in a vertical device. The ESD robustness of vertical LEDs with and without the internal protection diode was evaluated by reverse-bias ESD tests. The LEDs without the internal diode were found to have a pass yield of 0% at reverse voltages higher than 400 V; on the other hand, the LEDs with the internal diode showed a yield of 90% in the range of 2–4 kV.

Another approach consists of using a modulation-doped AlGaN-GaN superlattice structure, to improve the ESD reliability of nitride-based LEDs [22]. This approach can effectively improve the spreading of ESD pulse current, thus guaranteeing an ESD failure threshold in excess of 2000 V. On the other hand, Hwang et al. [51] proposed to insert a floating metal ring near the n-electrode of the LEDs. This floating metal reduces the peak electric field in the p–n junction via electrostatic charge induction, thus leading to a significant improvement of the ESD performance of the devices. Jang et al. [23] demonstrated that the thickness of the AlGaN electron blocking layer (EBL) can significantly impact on the ESD robustness of GaN LEDs. The increased robustness observed with thicker EBL was ascribed to the fact that the thickened EBL may partly fill the pits at the surface of the MQW region, which are due to the strain and low-temperature growth process. When these pits are not suppressed, they can reduce the ESD robustness of the devices.

Chiang et al. [27] investigated the impact of varied undoped GaN thickness on the optical performance and ESD robustness of LEDs. They demonstrated that as the thickness of the undoped GaN layer is increased, the density of V-shaped pits in the active region is suppressed. On the other hand, they demonstrated that the survival rate of LEDs with undoped GaN thickness of 1.5, 4, and 6 μm was 75%, 65%, and 55%: Devices with thin GaN have better ESD endurance than the other analyzed devices.

Recently, Chen [25] demonstrated that there is a strong correlation between the ESD stability and the structure of the electrodes of GaN-based LEDs. He proposed to use long parallel extension of the p- and n-side electrode areas to ease the distribution of the ESD current and enhance the robustness of the devices. This approach allows to improve significantly the ESD stability (+171% in anti-forward, +29% in anti-reverse) without adding more layers or altering the process procedures. Based on these results, Chen concluded that the uneven electric field is a key factor that strongly influences the ESD robustness of LEDs.

7.3 CONCLUSIONS

In this chapter, we have reviewed the main issues related to the ESD failure of electronic and optoelectronic devices based on GaN; after summarizing the main parameters of GaN, we have described the typical structure of transistors and LEDs based on this material and the related ESD issues. Within the text, the ESD performance of GaN-based devices has been compared to that of transistors and diodes based on more conventional compound semiconductors, such as GaAs and AlInGaP, to provide an extensive description of the topic.

The results summarized within this paper demonstrate that—even if GaN technology is reaching a certain degree of maturity—several processes may limit the ESD robustness of transistors and diodes based on gallium nitride; the important contribution of material defects (such as threading dislocations) and device layout to the overall ESD sensitivity has been extensively discussed within the text. Finally, we have described the problems related to the design of ESD protection structures based on GaN and presented the most promising approaches that have been recently proposed for the ESD protection of HEMTs and LEDs.

REFERENCES

1. A. Tazzoli, F. Danesin, E. Zanoni, and G. Meneghesso, "ESD robustness of AlGaN/GaN HEMT devices," in *Proceedings of the EOS/ESD Symposium*, 2007, 4A.4-1.
2. J. Kuzmík, D. Pogany, E. Gornik, P. Javorka, and P. Kordos, "Electrical overstress in AlGaN/GaN HEMTs: Study of degradation processes," *Solid-State Electronics*, 48, 271–276, 2004.
3. D. Cullen, D. Smith, A. Passaseo, V. Tasco, A. Stocco, M. Meneghini, G. Meneghesso, and E. Zanoni, "Electroluminescence and transmission electron microscopy characterization of reverse-biased AlGaN/GaN devices," *IEEE Transactions on Device and Materials Reliability*, 13, 126, 2013.
4. J. Joh and J.A. del Alamo, "Mechanisms for electrical degradation of GaN high-electron mobility transistors," in *IEEE IEDM Technical Digest*, 2006.
5. S. Bychikhin, R. Ferreyra, C. Ostermaier et al., "Investigation of nanosecond-time-scale dynamics of electric field distribution and breakdown phenomena in InAlN/GaN TLM structures," in *Proceedings of the European Workshop on Heterostructure Technology*, 2009.

6. S.H. Chen, A. Griffoni, P. Srivastava et al., "HBM ESD robustness of GaN-on-GaN-on-Si Schottky diodes," *IEEE Transactions on Device and Materials Reliability*, 12, 589, 2012.
7. Z. Wang, J.J. Liou, K.-L. Cho, and H.-C. Chiu, "Development of an electrostatic discharge protection solution in GaN technology," *IEEE Electron Device Letters*, 34, 1491, 2013.
8. M. Dal Lago, M. Meneghini, C. De Santi, M. Barbato, N. Trivellin, G. Meneghesso, and E. Zanoni, "ESD on GaN-based LEDs: An analysis based on dynamic electroluminescence measurements and current waveforms," *Microelectronics Reliability*, 54, 2138–2141, 2014.
9. G. Meneghesso, A. Chini, A. Maschietto, E. Zanoni, P. Malberti, and M. Ciappa, "Electrostatic discharge and electrical overstress on GaN/InGaN light emitting diodes," in *Proceedings of the EOS/ESD Symposium*, September 11–13, 2001, pp. 247, 252.
10. M. Meneghini, S. Vaccari, M. Dal Lago et al. "ESD degradation and robustness of RGB LEDs and modules: An investigation based on combined electrical and optical measurements," *Microelectronics Reliability*, 54, 1143, 2014.
11. M. Meneghini, A. Tazzoli, G. Mura, G. Meneghesso, and E. Zanoni, "A review on the physical mechanisms that limit the reliability of GaN-based LEDs," *IEEE Transactions on Electron Devices*, 57, 1, 108–118, 2010.
12. S. Vaccari, M. Meneghini, A. Griffoni, D. Barbisan, M. Barbato, S. Carraro, M. La Grassa, G. Meneghesso, and E. Zanoni, "ESD characterization of multi-chip RGB LEDs," *Microelectronics Reliability*, 53, 1510–1513, 2013.
13. C.A. Abboud, M. Chahine, C. Moussa, H.Y. Kanaan, and E.A. Rachid, "Modern power switches: The gallium nitride," in *Industrial Electronics and Applications, 2014 IEEE 9th Conference*, June 9–11, 2014, pp. 2203, 2208.
14. http://www.ioffe.rssi.ru/SVA/NSM/Semicond/GaN/electric.html.
15. E.F. Schubert, *Light-Emitting Diodes*, 2nd edition, Cambridge, 2006.
16. R.D. Dupuis and M.R. Krames, "History, development, and applications of high-brightness visible light-emitting diodes," *Journal Lightwave Technology*, 26, 9, 1154–1171, 2008.
17. G. Meneghesso, G. Verzellesi, F. Danesin et al., "Reliability of GaN high-electron-mobility transistors: State of the art and perspectives," *IEEE Transactions on Device and Materials Reliability*, 8, 2, 332–343, June 2008.
18. J. Kuzmk, D. Pogany, E. Gornik, P. Javorka, and P. Kordoš, "Electrostatic discharge effects in AlGaN/GaN high-electron-mobility transistors," *Applied Physics Letters*, 83, 4655, 2003.
19. S.-C. Lee, J.-C.L. Her, S.-M. Han, K.-S. Seo, and M.-K. Han, "Electrostatic discharge effects on AlGaN/GaN high electron mobility transistors," *Japanese Journal of Applied Physics*, 43, 4B, 2004, 1941–1943.
20. D.A. Gajewski, S. Sheppard, T. McNulty, J.B. Barner, J. Milligan, and J. Palmour, "Reliability of GaN/AlGaN HEMT MMIC Technology on 100 mm 4H-SiC," in *Proceedings of the 26th Annual JEDEC ROCS Workshop*, Indian Wells, CA, 2011.
21. C. Fleury, R. Zhytnytska, S. Bychikhin et al., "Statistics and localisation of vertical breakdown in AlGaN/GaN HEMTs on SiC and Si substrates for power applications," *Microelectronics Reliability*, 53, 1444–1449, 2013.
22. H.C. Chiu, K.-L Cho, and S.W. Peng, "A high protection voltage dual-gate GaN HEMT clamp for electric vehicle application," in *Electron Devices and Solid-State Circuits, 2011 International Conference*, November 17–18, 2011, pp. 1, 2.
23. M. Meneghini, G. Meneghesso, and E. Zanoni, "Electrical properties, reliability issues, and ESD robustness of InGaN-based LEDs," in *III-Nitride Based Light Emitting Diodes and Applications*, Springer, Dordrecht, the Netherlands.
24. T.C. Wen, S.J. Chang, C.T. Lee, W.C. Lai, and J.K. Sheu, "Nitride-based LEDs with modulation-doped Al Ga N–GaN superlattice structures," *IEEE Transaction on Electron Devices*, 51, 1743, 2004.

25. C. Jangm, J.K. Sheu, C.M. Tsai, S.C. Shei, W.C. Lai, and S.J. Lang, "Effect of thickness of the p-AlGaN electron blocking layer on the improvement of ESD characteristics in GaN-based LEDs," *IEEE Photonics Technology Letters*, 20, 1142, 2008.

26. R.W. Chuang, P.C. Tsai, Y.K. Su, and C.H. Chu, "Improved ESD properties by combining GaN-based light-emitting diode with MOS capacitor," *Solid-State Electronics*, 52, 1043–1046, 2008.

27. S.L. Chen, "Enhanced electrostatic discharge reliability in GaN-based light-emitting diodes by the electrode engineering," *IEEE Journal of Display Technology*, 10, 807, 2014.

28. H.H. Jeong, S.Y. Lee, J.H. Bae et al., "Improved electrostatic discharge protection in GaN-based vertical light-emitting diodes by an internal diode," *IEEE Photonics Technology Letters*, 23, 423, 2011.

29. T.H. Chiang, C.K. Wang, S.J. Chang et al., "Effect of varied undoped GaN thickness on ESD and optical properties of GaN-based LEDs," *IEEE Photonics Technology Letters*, 24, 800, 2012.

30. Y.J. Liu, D.F. Guo, L.Y. Chen et al., "Investigation of the electrostatic discharge performance of GaN-based light-emitting diodes with naturally textured p-GaN contact layers grown on Miscut Sapphire substrates," *IEEE Transaction on Electron Devices*, 57, 2155, 2010.

31. D. Bisi, M. Meneghini, C. de Santi et al., "Deep-level characterization in GaN HEMTs-Part I: Advantages and limitations of drain current transient measurements," *IEEE Transactions on Electron Devices*, 60, 3166, 2013.

32. R. Dietrich, A. Wieszt, A. Vescan, H. Leier, R. Stenzel, and W. Klix, "Power handling limits and degradation of large area AlGaN/GaN RF-HEMTs," *Solid-State Electronics*, 47, 123, 2002.

33. R. Vetury, N.-Q. Zhang, S. Keller, and U.K. Mishra, "The impact of surface states on the DC and RF characteristics of AlGaN/GaN HFETs," *IEEE Transactions on Electron Devices*, 48, 3, 560–566, 2001.

34. P. Srivastava, J. Das, and D. Visalli et al., "Silicon substrate removal of GaN DHFETs for enhanced (<1100 V) breakdown voltage," *IEEE Electron Device Letters*, 31, 851, 2010.

35. N. Herbecq, I. Roch-Jeune, N. Rolland, D. Visalli, J. Derluyn, S. Degroote, M. Germain, and F. Medjdoub, "1900 V, 1.6 m Ω cm^2 AlN/GaN-on-Si power devices realized by local substrate removal," *Applied Physics Express*, 7, 034103, 2014.

36. G.A. Ellis, J.S. Moon, D. Wong et al., "Wideband AlGaN/GaN HEMT MMIC low noise amplifier," in *IEEE MTT-S International Microwave Symposium Digest*, June 2004, pp. 153–156.

37. Aethercomm, *Gallium Nitride (GaN) Microwave Transistor Technology for Radar Applications*, http://www.aethercomm.com/articles/9.pdf.

38. E. Zanoni, M. Meneghini, A. Chini, D. Marcon, and G. Meneghesso, "AlGaN/GaN-based HEMTs failure physics and reliability: Mechanisms affecting gate edge and Schottky junction," *IEEE Transactions on Electron Devices*, 60, 10, 3119–3131, 2013.

39. J.L. Hudgins, G.S. Simin, E. Santi, and M.A. Khan, "An assessment of wide bandgap semiconductors for power devices," *IEEE Transactions on Power Electronics*, 18, 3, 907–914, May 2003.

40. K. Chen and C. Zhou, "Enhancement-mode AlGaN/GaN HEMT and MISHEMT technology," *Physica Status Solidi A*, 208, 2, 434–438, 2011.

41. E. Zanoni, F. Danesin, M. Meneghini, A. Cetronio, C. Lanzieri, M. Peroni, and G. Meneghesso, "Localized damage in AlGaN/GaN HEMTs induced by reverse-bias testing," *IEEE Electron Device Letters*, 30, 5, 427–429, 2009.

42. M. Meneghini, N. Trivellin, M. Pavesi, M. Manfredi, U. Zehnder, B. Hahn, G. Meneghesso, and E. Zanoni, "Leakage current and reverse-bias luminescence in InGaN-based light-emitting diodes," *Applied Physics Letters*, 95, 173507, 2009.

43. M. Meneghini, A. Tazzoli, R. Butendeich, B. Hahn, G. Meneghesso, and E. Zanoni, "Soft and hard failures of InGaN-based LEDs submitted to electrostatic discharge testing," *IEEE Electron Device Letters*, 31, 6, 579–581, 2010.

44. M. Meneghini, D. Bisi, D. Marcon, S. Van Hove, T.-L. Wu, S. Decoutere, G. Meneghesso, and E. Zanoni, "Trapping and reliability assessment in D-Mode GaN-based MIS-HEMTs for power applications," *IEEE Transactions on Power Electronics*, 29, 2199, 2014.

45. K. Bock, "ESD issues in compound semiconductor high-frequency devices and circuits," *Microelectronics Reliability*, 38, 1781, 1998.

46. K. Bock and H.L. Hartnagel, "Review: Surface technology and ESD protection towards highly reliable GaAs microwave circuits," *IOP Semiconductor Science and Technology*, 9, 1005, 1994.

47. A. Christou, editor, *Reliability of Gallium Arsenide MMICs*, Wiley, Chichester, West Sussex, 1992.

48. F.A. Buot and K.J. Sleger, "Numerical simulation of hot electron effects in GaAs FETs and MODFETs," *Solid State Electronics*, 27, 1067, 1987.

49. W.T. Anderson Jr., and E.W. Chase, "Electrostatic discharge effects in GaAs FETs and MODFETs," *EOS/ESD-Symposium Proceedings*, ESD Association, 1987, 205.

50. Y. Ota, M. Yanagihara, and M. Nakamura, *Solid State Electronics*, 38, 2005, 1994.

51. http://www.cree.com/xlamp_app_notes/electrical_overstress.

52. S.J. Chang, C.H. Chen, Y.K. Su et al., "Improved ESD protection by combining InGaN–GaN MQW LEDs with GaN Schottky diodes," *IEEE Electron Device Letters*, 24, 129, 2003.

53. S. Hwang and J. Shim, "Improved ESD voltage by inserting floating metal ring in GaN-based light emitting diodes," *Electronics Letters*, 44, 590, 2008.

8 ESD Protection Circuits Using NMOS Parasitic Bipolar Transistor

Teruo Suzuki

CONTENTS

8.1 INTRODUCTION

This chapter describes the case when an NMOS parasitic bipolar transistor is used in electrostatic discharge (ESD) protection circuit. The ESD protection circuit structure is shown in Figure 8.1. ESD stress application is divided into the following two types:

1. Apply ESD to signal pin. Reference is power supply pin or GND pin.
2. Apply ESD between power supply pin and GND pin (power supply ESD).

Therefore, ESD protection circuits need to be placed between signal and power/GND and between power supply and GND. Power supply ESD described in point 2 means ESD is applied between power supply pin and GND pin. (One of them is the pin that ESD is applied. The other is the reference pin.) ESD protection circuit for power supply and GND (power clamp) protects internal circuit (protected circuit) from ESD zapping. ESD protection circuit of signal area uses NMOS between signal and V_{ss} and PMOS between signal and power supply. Power rail clamp uses NMOS similar

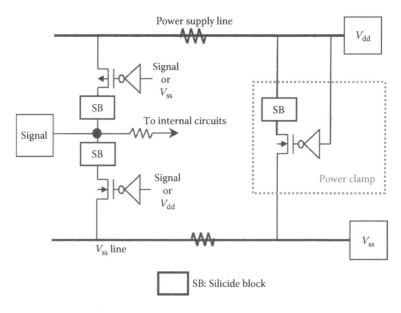

FIGURE 8.1   ESD protection circuit with MOS transistor.

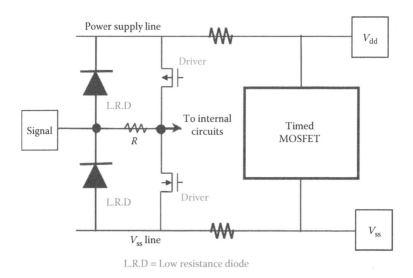

FIGURE 8.2   ESD protection circuit with dual-diode and timed MOSFET.

to ESD protection circuit between signal and V_{ss}. However, the transistor size of the power clamp is at least two times bigger. At present, most common ESD protection network uses dual-diode and timed MOSFET, as shown in Figure 8.2. However, in signal I/O, NMOS parasitic bipolar transistor ESD protection circuit is needed especially for NMOS open drain structure I/O cell-like inter-integrated circuit (I2C).

In power clamp, NMOS parasitic bipolar-type ESD protection circuit is needed for very fast power ramp-up case to avoid inrush current of timed MOSFET.

8.2 ESD PROTECTION BY NMOS

This section describes NMOS ESD protection circuit using circuit diagram, sectional view, and the current-voltage (I–V) characteristic in Figure 8.3. V_{ss} (–) means application of a negative voltage (–) to signal pin with reference to V_{ss}. By V_{ss} (–), current flows from P_{sub} (V_{ss}) to n$^+$ (signal pin). Forward direction diodes allow the flow of current over the entire P–N junction area, avoiding generation of local hot spots, so these are robust ESD protection elements. V_{ss} (+) means application of a positive voltage (+) to signal pin with reference to V_{ss}. By V_{ss} (+), current is flown by parasitic lateral NPN (LNPN) bipolar transistors. When the drain voltage increases, the following occurs:

1. Flow of current occurs caused by avalanche breakdown between the drain and substrate.
2. The potential of substrate rises locally, causing a potential difference of 0.6 V between the base and emitter.
3. The LNPN is turned on and large current (collector current) flows.
4. As a high electric field is applied between the drain and substrate, ESD damage often occurs between the drain and substrate [1].

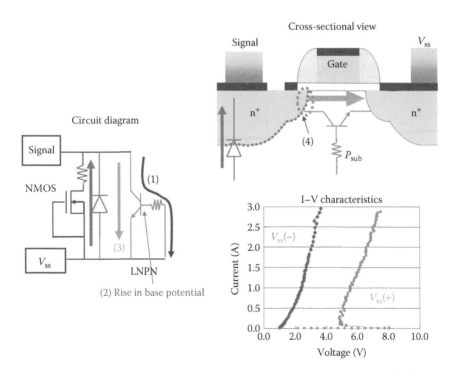

FIGURE 8.3 Circuit diagram, cross-sectional view, and I–V characteristics of NMOS.

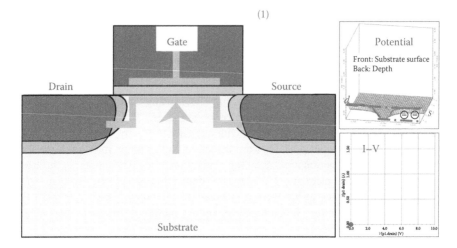

FIGURE 8.4   Normal condition.

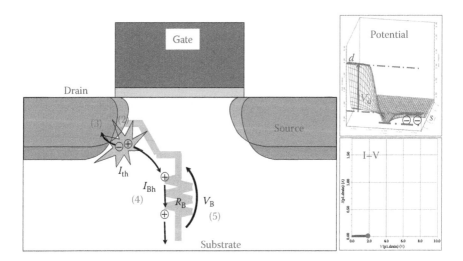

FIGURE 8.5 $V_d < V_{t1}$.

How NMOS parasitic bipolar electricity discharges ESD surge is explained in detail using sectional view, potential, and I–V characteristic from Figures 8.4 through 8.9.

1. NMOS is OFF.
2. Because of electric field, weak avalanche multiplication (collisional ionization) occurs and generates electron–hole pairs.
3. Electrons flow to drain.
4. Holes ($I_{Bh} = [M - 1] \times I_{th}$) flow to substrate (where M is avalanche multiplication coefficient).
5. Base voltage is gradually increased by $V_B = I_{bh} \times R_B$.

6. Base voltage V_B increases up to approx. 0.6 V.
7. E–B junction barrier disappears (biased) and electrons are implanted from emitter.
8. Bipolar discharge by I_{CE} ($= I_{EE}$) starts.
9. $I_{Bh} = (M - 1) \times (I_{CE} + I_{th})$ and I_{Bh} is supplied regardless of small M. V_d becomes smaller and leads to snapback condition.
10. At V_h, bipolar is completely ON.
11. V_d (V_{CE}) is increased again by ON current increase.

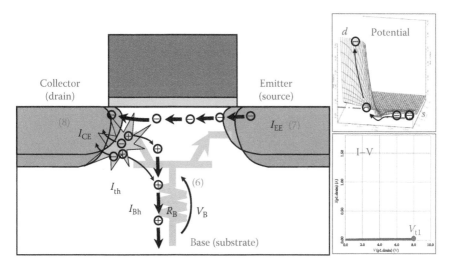

FIGURE 8.6 $V_d = V_{t1}$.

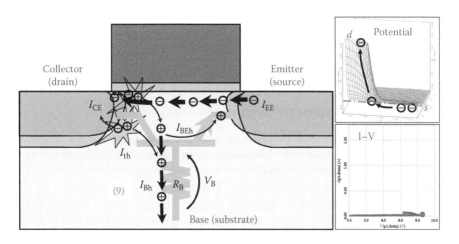

FIGURE 8.7 $V_{t1} < V_d < V_h$.

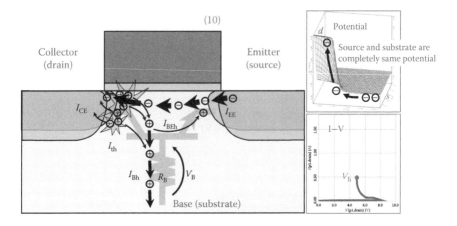

FIGURE 8.8 $V_d = V_{hold}$.

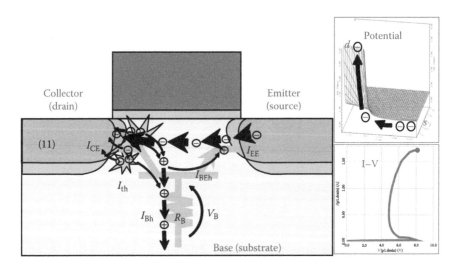

FIGURE 8.9 $V_d > V_{hold}$.

8.3 ESD PROTECTION BY PMOS

This section describes PMOS ESD protection circuit using circuit diagram, cross-sectional view, and I–V characteristic in Figure 8.10. Compared with Figure 8.3, the structure is similar, but P-type and N-type are reversed. V_{dd} (+) means application of a positive voltage (+) to signal pin with reference to V_{dd}. By V_{dd} (+), current flows from P+ (signal pin) to N-well (power supply). Forward direction diodes allow the flow of large current. V_{dd} (−) means application of a negative voltage (−) to signal pin with reference to V_{dd}. By V_{dd} (−), current flows in parasitic lateral PNP (LPNP) bipolar transistors. In LPNP bipolar transistors, where many

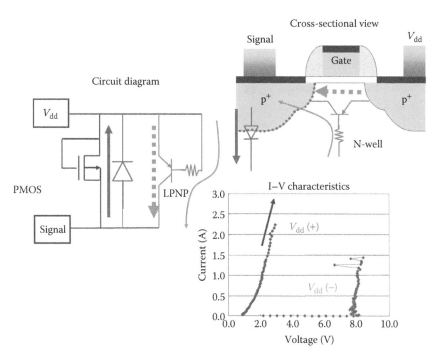

FIGURE 8.10   Circuit diagram, cross-sectional view, and I–V characteristics of PMOS.

charge carriers are holes, the β (hfe) of parasitic bipolar has a lower value com-pared with that in NMOS, so current flow is smaller. Therefore, with PMOS, the width of transistors needs to be larger [2,3].

8.4   BALLAST RESISTORS

For high-speed semiconductor device, drain, gate, and source are metallized using $TiSi_2$, $CoSi_2$, and NiSi. Such a structure is said to be silicided. However, a silicided structure is weak against ESD because of shallow junction depth of drain and source [4]. Because of this shallow junction, LNPN emitter efficiency is degraded, which causes lower ESD robustness [5,6]. If emitter efficiency is low, substrate current increases to keep parasitic bipolar transistor of ESD protection circuit switched ON. As a result, power consumption increases and ESD robustness decreases. Generally, the transistor width of ESD protection circuit is too large to lay out as one transistor. Therefore, the transistor is divided into uniform width transistors, as shown in Figure 8.11. As each divided transistor look like a finger, this is called a multi-finger transistor circuit. In a silicided structure, the drain resistance is very small. Therefore, not all parallel ESD protection circuits work equally when ESD is applied. As a result, cur-rent is concentrated in certain protection circuits and ESD robustness is decreased [7]. To avoid such condition, resistance is added on the drain side, which is referred to as ballast resistance. When one of the parasitic LNPN bipolar transistors of a finger

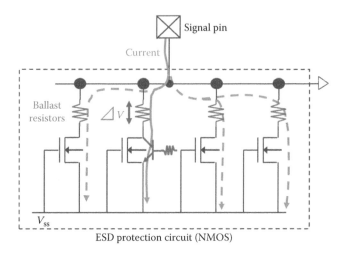

FIGURE 8.11   Ballast resistors.

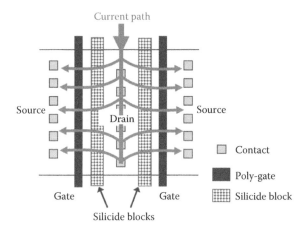

FIGURE 8.12   Silicide block.

transistor turns on, the pad potential decreases. However, inserting ballast resistors between the pad and drain, voltage drop of the pad potential is controlled and make it easier to turn on the entire LNPN. The ballast resistance that is often generally used is a silicide block as shown in Figure 8.12 [8–10], where a higher resistance non-silicided partof the drain acts as a ballast resistor.

8.5   ESD DESIGN WINDOW

Figure 8.13 shows an ESD design window [11]. ESD design window means the area inside limit lines (1), (2), (3), and the X-axis (voltage). Limit lines (1), (2), and (3) are defined as follows:

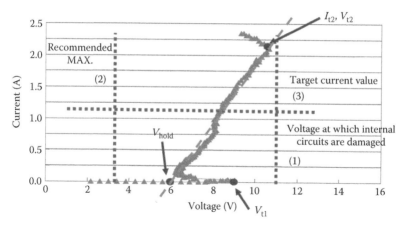

FIGURE 8.13   ESD design window.

1. *Failure voltage of the internal circuits (protected circuits)*
   ESD protection circuit should work at lower voltage than this failure voltage, otherwise the internal circuit is damaged. Failure voltage of the internal circuit is decided by lowest failure voltage of gate oxidation film or the inverter with the unstable gate electric potential [12]. In general, failure voltage of the internal circuit is tested by transmission line pulse (TLP) [13–17], which outputs ESD pulse-like voltage waveform, because ESD is a high-frequency discharge phenomenon, and failure voltage by DC measurement is lower than actual peak voltage. Because of technology scaling, the failure voltage of internal circuit is getting lower and lower, and ESD protection circuit design is becoming more difficult [18].

2. *The maximum rating in normal use*
   V_{hold} (hold voltage) should be higher than the maximum rating in normal use, because the ESD protection circuit (parasitic bipolar shown in Figure 8.13) has to avoid being turned on in normal chip operation.

3. *Sufficient ESD current*
   The second breakdown voltage is called V_{t2} with current I_{t2}. I_{t2} is the reference ESD robustness, which is good when I_{t2} is high [19]. In ESD design, it is necessary to conduct the ESD current with lower voltage than the internal circuit failure voltage. In Figure 8.13, I_{t2} is 2.4 A. Discharge current was 1.33 A when HBM = 2000 V is applied. Therefore, discharge performance of ESD protection circuit is sufficient. The voltage when conducting 1.33 A is 8 V. As the voltage is lower than internal circuit failure voltage (11 V), ESD protection circuit is sufficient. As explained in (1), (2), and (3), it is important to discharge sufficient ESD current before the internal circuit failure to satisfy the ESD standard. Important points of ESD design are as follows:
   a. Understand the discharge capability of the ESD protection circuit.
   b. Provide low-resistance discharge paths for ESD surge current.
   c. Before starting the ESD design, check failure voltage of protected circuits.

8.6   ESD CHARACTERISTIC WITH P/P⁺ EPITAXIAL SUBSTRATE

During the ten years from 1995, epitaxial substrate (Epi) of P/P⁺ type was used for mass production of MOS-type semiconductor device. Even very small defects in the silicon crystal are not negligible because of the large-scale integration (LSI) scaling, so it is necessary to use low-defect epitaxial wafers to improve yield. However, as noted in Section 8.2, it is better to use high substrate resistance for NMOS parasitic bipolar characteristics. In P/P⁺ Epi substrate, ESD robustness is weak because substrate resistance is very low [20]. P/P⁺ Epi substrate is a wafer where an Epi layer of specific resistance of 10 Ω cm is grown on the low-specific resistance substrate of 0.01–0.02 Ω cm, as shown in Figure 8.14. The thickness of the Epi layer is usually around 5–10 μm.

The bulk resistance of a conventional Czochralski method (CZ) substrate is about 10 Ω cm, the same as an Epi layer. Figure 8.15 shows the snapback characteristics of three types of Epi substrate. Epi film thickness 5 μm: Epi5 and thickness 10 μm: Epi10 and a CZ substrate [21]. The snapback voltage of these three substrates are all about 12 V. However, LNPN turn-on current (I_{t1}) is 3 mA in CZ substrate, 10 mA in Epi10 substrate, and 30 mA in Epi5. The resistance value calculated from waveform slope at breakdown is about 20 Ω for the Epi5, about 60 Ω for the Epi10, and about 200 Ω for the CZ. The voltage between base and emitter to turn on the LNPN transistor is about 0.6 V in all substrates. When a specific resistance of a substrate is low, it is necessary to flow big currents into the substrate to turn on the LNPN. From test result, required current for each substrate type is Epi5 > Epi10 > CZ. With higher current, the protection circuit does not turn on smoothly, heat generation increases, and ESD becomes weak. To avoid this issue, it is necessary to increase the substrate resistance to turn on ESD protection circuit smoothly.

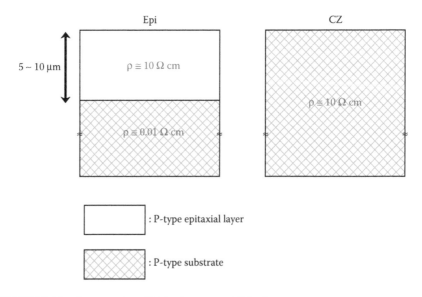

FIGURE 8.14   Cross section of Epi substrate and CZ substrate.

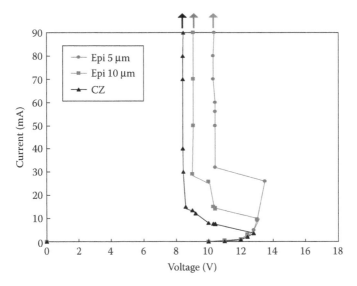

FIGURE 8.15   Snapback characteristics of Epi5, Epi10, and CZ.

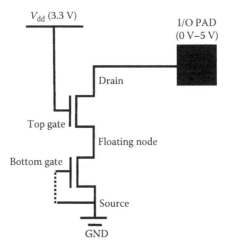

FIGURE 8.16   Circuit of 5 V-tolerant I/O cell.

8.7   ESD IMPROVEMENT BY ADDITIONAL BORON (B⁺) ION IMPLANTATION

As written in Section 8.5, it is important to understand ESD design window for ESD protection design. Internal circuit failure voltage becomes lower and lower because of technology scaling, and ESD protection becomes more difficult. Therefore, it is important to find way to lower V_{t1} of the NMOS parasitic bipolar transistor.

One method to lower V_{t1} is gate bias and back gate bias [9]. The other method is make a steep P–N junction. The method to make steep P–N junction is B$^+$ (boron) ion implantation to drain [22–26]. This section describes method to improve ESD robustness by B$^+$ ion implantation under the drain contact area without degrading the AC characteristics of the circuit.

8.7.1 ESD Improvement for Overvoltage Tolerant

Because of technology scaling, the power supply voltage of semiconductor device has been decreasing over time. However, as 5 V power has been used quite a long time, 5 V LSI are still used on mother board. As a result, 3.3 V LSI needs I/O, which can accept 5 V input. This kind of I/O cell is generally called an "overvoltage-tolerant I/O cell." A 5 V-tolerant I/O cell that accept 5 V signal by using 3.3 V power supply uses cascaded connection transistor to secure transistor reliability, as shown in Figures 8.16 and 8.17. However, cascaded structure NMOS transistor's ESD protection performance is lower compared with the single NMOS structure. The reason is that the effective base length of the LNPN parasitic bipolar transistor becomes long [27,28].

To investigate how V_{t1} decrease by P–N junction slope change, I simulated I–V characteristics applying HBM = 3000 V by process simulator "T-SUPREM4" and device simulator "MEDICI."

Figure 8.18 shows TCAD (using MEDICI) simulation result of HBM = 3000 V.

Solid line indicates reference. The rough broken line indicates doubled NMOS L_{dd} energy dosage where R_{ON} became a little steeper. The broken line of moderate density indicates 10 times L_{dd} dosage amount. In this case, V_{hold} (V_{t1}) was decreased and R_{ON} became steep. However, this method changes the transistor characteristics.

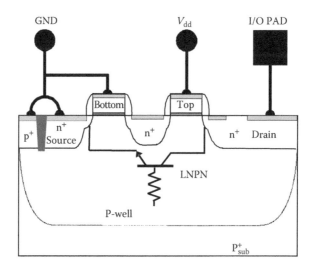

FIGURE 8.17 Cross section of 5 V-tolerant I/O cell.

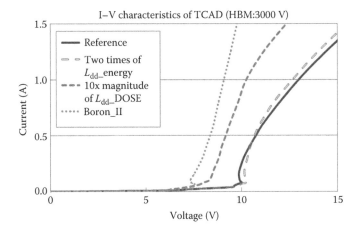

FIGURE 8.18 TCAD simulation results.

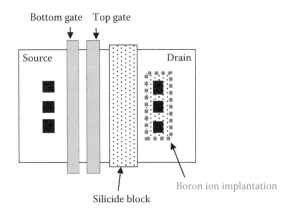

FIGURE 8.19 Additional boron ion implantation.

Therefore, transistor reliability test is necessary for the structure so it has a cost disadvantage, and HCI (which is trade-off relation) becomes bad. The small broken line indicates additional boron ion implantation of Figures 8.19 and 8.20.

The boron ion implantation area is outside the poly-gate region, so it does not change transistor characteristics. From business point of view, it has cost advantage because additional reliability evaluation such as HCI is not necessary. Therefore, this ion implant method can save test time.

As ion implant area is outside of silicide block area, there is no effect for silicide block resistance. So this method improves ESD robustness of small amplitude IP-like USB2.0 without changing AC characteristics.

Also this method can be used to improve ESD robustness even when product is in mass production stage.

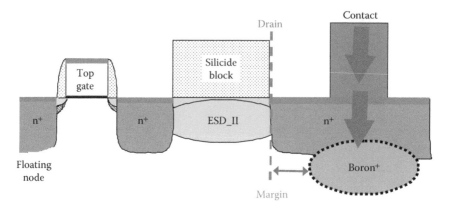

FIGURE 8.20   Additional boron ion implantation (cross section).

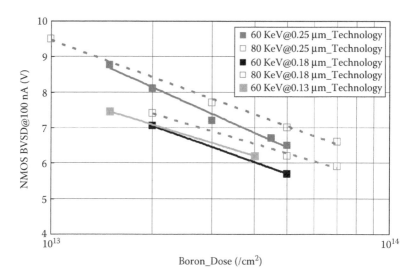

FIGURE 8.21   The relationship of boron dose and BVSD among the three technology.

To save process step and cost, the silicide block was made with the same process step as poly-gate sidewall in Figure 8.20. Therefore, in n+ drains and sources area, there is no ion implantation under the silicide block. To lower the resistance under the silicide block, P (phosphorus) was implanted.

Figure 8.21 shows relation between boron ion implantation amount and BVSD in three technology nodes (0.25 μm, 0.18 μm, and 130 nm technology). Suppose appropriate BVSD is around 7 V, optimum amount of boron ion implantation is in 2E13–5E13 range. This method was effective in all three technology nodes.

8.7.2 Application to Single Structure

Figure 8.22 shows I–V characteristics of single-structure NMOS transistor with B^+ ion implantation.

Table 8.1 shows ESD parameter (V_{t1}, V_{hold}, I_{t2}) of I–V characteristics. V_{t1} becomes 1.2 V lower. I_{t2} is improved by 50%. As a result, B^+ ion implantation is effective for ESD discharge performance improvement of single-structure NMOS transistor as well as cascaded structure NMOS transistor.

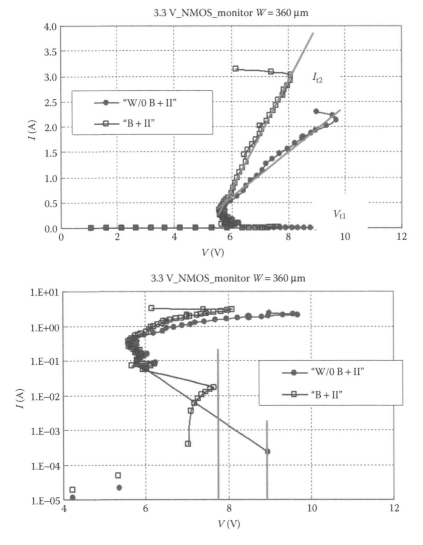

FIGURE 8.22 I–V characteristics after additional boron ion implantation for single NMOS transistor. (Top: The vertical axis is in linear scale. Bottom: The vertical axis is in log scale.)

TABLE 8.1

ESD Parameters from Figure 8.22

	V_{t1} (V)	V_h (V)	I_{t2} (A)
W/O B+II	8.9	5.8	2.1
B+II	7.7	5.7	3.0

8.8 EFFECT OF POWER–GND CAPACITANCE FOR NMOS PROTECTION CIRCUIT

When using NMOS parasitic bipolar as power clamp, one factor that affects the ESD protection characteristics is the parasitic capacitance between power and GND [29].

As shown in Figure 8.23, when ESD test of device I performed using V_{ss} as reference and surge ESD to power (V_{dd}) pin, the parasitic capacitance of the whole device can discharge the ESD surge. Therefore, a bigger chip has better ESD robustness.

However, recent LSI designs require multi-V_{dd} supplies to manage the power consumption or to decrease the noise propagation from a digital block to a precision analog block. When power is isolated, capacitance between power (V_{dd}) and V_{ss} becomes smaller. Therefore, ESD through parasitic capacitance becomes smaller.

The ESD designer needs to pay attention to this power supply isolation. However, it was found that the ESD robustness is not simply proportional to the size of the parasitic capacitance between the supply lines.

The power supply isolation is divided into three types by the number of I/O cells. It is defined as small scale (less than five I/O cells), medium scale (less than 100 I/O cells), and large scale (more than 100 I/O cells). Table 8.2 shows the measurement results.

In case of power isolation scale is medium, ESD robustness was MM = 50 V–200 V, HBM = 500 V–2000 V, which was worse than small-scale and large-scale cases.

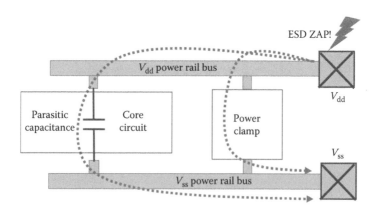

FIGURE 8.23 Example of power-supply ESD protection. This session studies the influence on the ESD behavior of the parasitic capacitance.

TABLE 8.2

ESD Measurement Results of HBM and MM

	Power Supply Separation (The Number of I/O Cell)		
	Small (Until 5)	Middle (Until 100)	Large (More Than 100)
HBM	>2000 V	500–2000 V	>2000 V
MM	>200 V	50–200 V	>200 V
Capacitance between supply lines	–	60–120 pF	–

The capacitance value in a medium-scale was 60–200 [pF] with LCR meter.

This result shows large-scale isolation does not always have the best ESD robustness.

The positive ESD stress is simulated for different values of parasitic capacitances from 10 pF to 10 nF. Two cases are simulated: gate-grounded NMOS (GGNMOS) and $V_g = 1$ V constant.

Figure 8.24 shows the result of GGNMOS, and Figure 8.25 shows the result of $V_g = 1$ V. In the case of GGNMOS, a peak current could be observed for all added capacitances of 10 pF, 100 pF, 1 nF, and 10 nF. The peak current becomes about 7 A for the 10 nF case. On the other hand, in the case of $V_g = 1$ V, the current for the 10 nF case is reduced to 3.5 A. It is expected that the electric charge in the capacitance between supply lines had decreased because the V_{t1} with $V_g = 1$ V

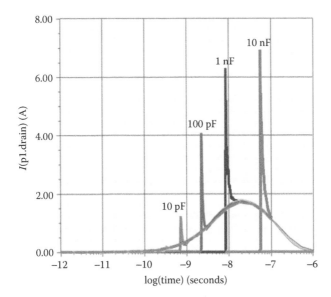

FIGURE 8.24 The current characteristics of GGNMOS. Capacitances added in parallel discharge additional current into the GGNMOS at snapback. The peak current can be very high.

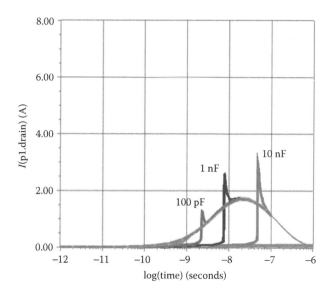

FIGURE 8.25 The current characteristics ($V_g = 1$ V).

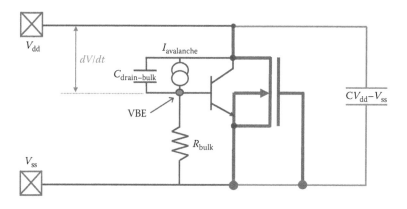

FIGURE 8.26 Parasitic capacitance of $C_{\text{drain–bulk}}$.

is reduced compared with the case with GGNMOS. Thus, the peak current is reduced.

The ESD robustness of the medium-scale versus the small-scale case is reviewed because of the supply capacitance using Figure 8.26. The rise of the applied ESD voltage is significantly slowed down when the capacitance $CV_{dd}–V_{ss}$ between V_{dd} and V_{ss} is about 100 pF, and the rise of the potential of V_{dd} slows. Therefore, the displacement current ($C_{\text{drain–bulk}} \times dV/dt$) decreases [30]. To turn on the LNPN power clamp, the potential of the base (VBE) should become 0.6 V. VBE is shown by the following formula:

$$\text{VBE} = R_{\text{bulk}} \times \left(I_{\text{avalanche}} + C_{\text{drain-bulk}} \times \frac{dV}{dt} \right) \qquad (8.1)$$

The avalanche current $I_{\text{avalanche}}$ supplements the drain-bulk displacement current to reach VBE = 0.6 V. It is found that the temperature increase at the P–N junction due to the avalanche current dissipation causes the ESD damage.

In addition, the displacement current through $CV_{\text{dd}}-V_{\text{ss}}$ consumes a higher amount of the ESD (source) stress current when this capacitance becomes bigger (10 pF–10 nF). However, the positive peak current through the LNPN also becomes higher (10 pF–10 nF) because the impedance between the $CV_{\text{dd}}-V_{\text{ss}}$ and LNPN is very low. This positive peak current is understood as the main cause of low ESD robustness for the medium-scale isolation.

This phenomenon is similar to the problem of test-board capacitance discharge at snapback [31,32]. The aforementioned drain-bulk displacement current grows for the small power supply isolation. Therefore, the thermal destruction of P–N junction by thermal heat does not occur easily because the avalanche current is required to turn on the LNPN decreases. When the capacitance of the power supply isolation is large and is about 50 nF, the ESD surge flows through the capacitance. Therefore, the voltage between supply lines does not become high enough to destroy the device.

8.9　SUMMARY

This chapter described ESD protection mechanism of NMOS parasitic NPN bipolar transistor. In ESD design, importance of ESD design window was described. The mechanism was explained how ESD robustness is degraded by decreasing Epi substrate resistance. To improve ESD robustness, it is effective to increase substrate resistance by reducing the doping concentration of the substrate under the Epi.

Next, this report described ESD robustness improvement method of cascaded NMOS transistor, which is used for overvoltage-tolerant I/O cell-like 5 V-tolerant I/O. By implanting B^+ under the drain contact, ESD robustness was improved. This method has the following advantages:

- Additional reliability test is not required because transistor characteristics is not changed.
- The method is applicable for product already in mass production stage.
- Improves ESD robustness without changing AC characteristics of USB2.0 or XTAL I/O.

Finally, it was shown that increased parasitic capacitance across LNPN devices does not necessarily enhance the ESD robustness.

When the power supply isolation is medium scale, it is found that the drain-bulk displacement current decreases, the LNPN avalanche trigger current increases, and the P–N junctions fail early.

REFERENCES

1. Y. Tajima and I. Shinohara, "ESD improvement of input and output protection circuits in CMOS LSI," in *RCJ EOS/ESD/EMC Symposium Proceedings*, pp. 13–20, 1993.
2. T. Suzuki, J. Iwahori, T. Morita, H. Mochizuki, and H. Takaoka, "The research on ESD destruction of PMOS in the advanced technology," in *RCJ EOS/ESD/EMC Symposium Proceedings*, pp. 205–208, 2003.
3. B. Gianluca, D. Charvaka, and R. Vijay, "Efficient pnp characteristics of pMOS transistors in sub-0.13 μm ESD protection circuits," in *EOS/ESD Symposium Proceedings*, pp. 260–269, 2002.
4. K.-L. Chen, "The effects of interconnect process and snapback voltage on the ESD failure threshold of NMOS transistors," in *EOS/ESD Symposium Proceedings*, pp. 212–219, 1988.
5. A. Amerasekera, C. Duvvury, V. Reddy, and M. Rodder, "Substrate triggering and salicide effects on ESD performance and protection circuit. Design in deep submicron CMOS process," in *IEDM Technical Digest*, pp. 547–550, 1995.
6. A. Amerasekera, V. Mcneil, and M. Rodder, "Correlating drain junction scaling, silicide thickness and lateral npn behavior with the EOS/ESD performance of a 0.25 μm CMOS process," in *IEDM Technical Digest*, pp. 893–896, 1996.
7. T. Polgreen and T. Chatterjee, "Improving the ESD failure threshold of silicided n-MOS output transistors by ensuring uniform current flow," in *EOS/ESD Symposium Proceedings*, pp. 167–174, 1989.
8. D. Krakauer and K. Mistry, "ESD protection in a 3.3 V sub-micron silicided CMOS technology," in *EOS/ESD Symposium of Proceedings*, pp. 250–257, 1992.
9. A. Amerasekera and C. Duvvury, *ESD in Silicon Integrated Circuits*, Second Edition, John Wiley & Sons, West Sussex, U.K., 2002.
10. C.H. Dias, S.M. Kang, and C. Duvvury, *Modeling of Electrical Overstress in Integrated Circuits*, Kluwer Academic Publishers, Norwell, MA, 1995.
11. M. Mergens, C. Russ, K. Verhaege, J. Armer, P. Jozwiak, and R. Mohn, "HHI-SCR for ESD protection and latch-up immune IC operation," in *EOS/ESD Symposium Proceedings*, pp. 10–17, 2002.
12. International ESD Workshop, *Consideration of Design Window for ESD Power-Clamp in Next Generation Devices*, MIRAI-Selete, Japan, 2010.
13. T. Maloney and N. Khurana, "Transmission line pulsing techniques for circuit modeling of ESD phenomena," in *EOS/ESD Symposium Proceedings*, pp. 49–54, 1985.
14. J. Barth, K. Verhaege, L. Henry, and J. Richner, "TLP calibration, correlation, standards, and new techniques," in *EOS/ESD Symposium Proceedings*, pp. 85–96, 2000.
15. S.H. Voldman et al., "Standardization of the transmission line pulse (TLP) methodology for electrostatic discharge (ESD)," in *EOS/ESD Symposium Proceedings*, pp. 1–10, 2003.
16. M. Sawada, "Study of device waveform for TLP measurement," in *RCJ EOS/ESD/ EMC Symposium Proceedings*, pp. 185–190, 2004.
17. T. Suzuki and K. Shiraishi, "Examination of short calibration problem of transmission line pulse," *IEICE Electronics Express*, Vol. 10, No. 5, pp. 1–8, 2013.
18. T. Suzuki, "Trend of the component level ESD protection technology of the semiconductor product," *Journal of the Institute of Electrostatics Japan*, Vol. 36, No. 5, pp. 272–275, 2012.
19. A. Amerasekera, L.V. Roozendaal, J. Bruines, and F. Kuper, "Characterization and modeling of second breakdown in NMOST's for the extraction of ESD-related process and design parameters," *IEEE Transactions on Electron Devices*, Vol. 38, No. 9, pp. 2161–2168, 1991.
20. T. Suzuki et al., "ESD and latch-up characteristics of semiconductor device with thin epitaxial substrate," in *EOS/ESD Symposium Proceedings*, pp. 199–207, 1998.

21. T. Suzuki, O. Yoshiaki, S. Mitarai, S. Ito, and H. Monma, "A study of optimum ESD protection circuit with semiconductor device processed by CMOS 0.18 μm technology," in *RCJ EOS/ESD/EMC Symposium Proceedings*, pp.135–142, 1998.

22. "Electro-static discharge protection device and method for manufacturing electro-static discharge protection device," United States Patent Application Publication, US 2008/0211028 A1.

23. H. Tang, S. Chen, S. Liu, M. Lee, C. Liu, M. Wang, and M. Jeng, "ESD protection for the tolerant I/O circuits using PESD implantation," *Journal of Electrostatics*, Vol. 54, pp. 293–300, 2002.

24. G. Zimmermann, P. Reiss, W.T. Chang, H. Tang, H. Ou, and H. Cerva, "Dislocation induced leakage of P+ implanted ESD test macros in 90 nm technology," in *Proceedings of ISTFA*, pp. 120–125, 2004.

25. D. Alvarez, M.J. Abou-Khalil, C. Russ, K. Chatty, R. Gauthier, D. Kontos, J. Li, C. Seguin, and R. Halbach, "Analysis of ESD failure mechanism in 65 nm bulk CMOS ESD NMOSFETs with ESD implant," *Microelectronics Reliability*, Vol. 46, pp. 1597–1602, 2006.

26. K. Chatty, D. Alvarez, R. Gauthier, C. Russ, M. Abou-Khalil, and B.J. Kwon, "Process and design optimization of a protection scheme based on NMOSFETs with ESD implant in 65 nm and 45 nm CMOS technologies," in *EOS/ESD Symposium Proceedings*, pp. 385–394, 2007.

27. J.W. Miller, M.G. Khazhinsky, and J.C. Weldon, "Engineering the cascoded NMOS output buffer for maximum Vt1," in *EOS/ESD Symposium Proceedings*, pp. 308–317, 2000.

28. W.R. Anderson and D.B. Krakauer, "ESD protection for mixed-voltage I/O using NMOS transistors stacked in a cascode configuration," in *EOS/ESD symposium Proceedings*, pp. 54–62, 1998.

29. T. Suzuki, J. Iwahori, T. Morita, H. Takaoka, T. Nomura, K. Hashimoto, and S. Ichino, "A study of relation between a power supply ESD and parasitic capacitance," in *EOS/ESD Symposium Proceedings*, pp. 290–297, 2005.

30. B. Keppens, M. Mergens, J. Armer, P. Jozwiak, G. Taylor, R. Mohn, C. Trinh, C. Russ, K. Verhaege, and F. Ranter, "Active-area-segmentation (AAS) technique for compact ESD robust fully silicided NMOS design," in *EOS/ESD Symposium Proceedings*, pp. 250–258, 2003.

31. K. Verhage et al., "Analysis of HBM testers and specifications using a 4th order lumped element model," in *EOS/ESD Symposium Proceedings*, pp. 129–137, 1993.

32. C. Russ et al., "ESD protection elements during HBM stress tests—Further numerical and experimental results," in *EOS/ESD Symposium Proceedings*, pp. 97–105, 1994.

9 ESD Development in Foundry Processes

Jim Vinson

CONTENTS

9.1   INTRODUCTION

In modern society, electronic devices penetrate every area of our lives. Our day starts by waking up with the alarm on our smartphone. We get out of bed to fix a hot cup of coffee from a single-cup coffee brewing system. We proceed to read the morning paper on our tablet and also check our e-mail before heading out to work. Electronics are so much a part of our life that they are not even considered unique any longer. We are tied to our electronics. Today's generation has not experienced life without these modern conveniences of technology, and many do not know how to function without them. At the core of these modern devices is the integrated circuit (IC). The IC has allowed us to realize all these modern devices, but it is hidden from our view by the stylish and attractive enclosures manufactures package it in.

The modern IC is an amazing marvel of complexity. The electronics industry has been able to integrate billions of transistors into the modern microprocessor [1]. This level of integration allows unparalleled sophistication in the computing capability of the processor. The microprocessor is not the only component to benefit from these levels of integration. Power management components, which are less sexy than a microprocessor, have also seen a rise in complexity and integration especially in smartphones and tablets where space and battery life are at a premium.

A power management IC (PMIC) is an electronic system that converts the energy in a battery into the necessary power (voltage and current) for each subsystem. Each component in a smartphone must have power to operate correctly. The power must come in the form of a very stable voltage under varying load currents. A smartphone has a large number of subsystems, and each may have a different voltage and current requirement. The microprocessor may require one voltage, the display may require another voltage, and the RF transceivers yet another voltage. This places a big demand on the IC supplier to develop a circuit that can provide these varying voltages and power requirements all from a single battery source. This challenge is compounded by the reality that the battery voltage may be higher or lower than the desired voltage based on the charge state of the battery. The voltage may need to be reduced or boosted to meet the system requirements. As the desire is for long life and small size in our mobile devices, the power management circuits have to accomplish this conversion with very high efficiency (low energy loss) and extremely small physical size.

Designing these marvels of technology is not a trivial task. Teams of designers collaborate to produce the final design. One part of the design is the electrostatic discharge (ESD) protection. ESD protection is a necessary aspect of the design but does not add to the functional or performance specification. In fact, the ESD robustness of the IC may have to be balanced with the performance of the device. High ESD may mean lower performance, so to achieve the desired performance, ESD robustness may have to be sacrificed. All circuits need some ESD protection because without it the IC would be more prone to failure.

The final IC starts with an idea. The IC manufacturer defines a product they believe will be profitable and provide a good return on their investment. There are four main tasks that are needed to bring a new product to the marketplace. These are design, wafer fabrication, assembly, and test. Very large IC manufacturers have all four of these areas under their control. Smaller manufacturers may outsource one or more of these aspects to an outside vendor. Wafer fabrication, assembly, and test are the most common aspects outsourced. The design of a circuit is usually the key differentiator for a product if multiple manufactures have access to the same technology.

It is very costly for a manufacturer to own and operate the wafer fabrication, assembly, and test facilities. The start-up cost for each of these facilities is high, and they need to be running at near capacity to be cost-effective. Most IC suppliers cannot afford to do this. This is especially true for wafer fabrication. As a result, many IC manufacturers use outside wafer fabrication services (foundries) to build their products. This choice can complicate the development of a new product. The focus of this chapter is to detail the process used to implement a new IC design in a foundry process and also highlight the challenges with respect to ESD design.

9.2 PRODUCT DEFINITION

The concept for a new product originates from a perceived need in the marketplace. This could come from internal sources such as applications or marketing or from external sources such as customers or industry partners. The concept phase of the product defines a wish list of functions and performance specifications. During this stage of the process, market data will be gathered and estimated to develop a business plan for the product. This data, coupled with the estimated development cost and development timeline, will help the management team make an informed decision on whether it is financially beneficial to invest in this product. Companies are looking to maximize their profits, and much of the new money comes from design wins on new products. The preliminary timeline to develop the product is very important because it has a big influence on the revenue a new product can produce. The earlier a product is released and more innovative it is, the more revenue can be obtained prior to other companies releasing a competitive product. Misjudging the time to develop a new product may have disastrous effects on its profitability. The two factors of innovation and time to market are inversely correlated. Highly innovative products contain a lot of new to the company technology and maybe even new to the world technology. It is challenging for the development team to accomplish this level of innovation on an aggressive schedule. A balance of innovation and time to market need to be obtained. Schedule risk can also be reduced when multiple options are worked in parallel so the product does not have to rely on the success of a single path. This risk mitigation technique increases the development team size and also the development cost but can reduce the schedule risk associated with a very innovative product and improve first-pass success. One of the key things to consider at the start is to understand the technological and schedule risks associated with a project and work proactively to mitigate these risks.

9.2.1 Voltage Requirements

One of the first topics that must be agreed upon when defining a new product is its operating voltage or voltages. This set of specifications will dictate many aspects and constraints of the product. The voltage specification is not just a single number like 5 V. It is actually a group of numbers based on the operating conditions and environment of the part. Some parts may have multiple power domains, and therefore multiple sets of numbers are needed to properly define each power domain. There are three key values related to ESD that need to be considered for each voltage defined for the part. The first is the maximum operating voltage (MOV). The MOV can be defined as the upper end of a range such as 12 V when the operating range is specified as 5–12 V or the voltage plus a tolerance such as 5.5 V when the operating voltage is specified as 5 V $\pm$ 10%. The MOV is defined as the highest DC voltage level or peak voltage in a repetitive signal that can be applied to a device terminal for the duration of its life. This is the voltage used to define the device's long-term reliability (wear-out reliability). It is expected that the part is functional and meets all its expected parametric requirements. As noted, it could be the peak of a repetitive signal. In the case of switching devices, some signals have a DC component and an AC component riding on top of this DC. The MOV would be the DC + AC voltage, as shown in Figure 9.1. As MOV is tied to the operating reliability of the product, operation above MOV may degrade reliability. Operation above MOV is not recommended.

The second number of importance is the absolute maximum voltage (AMV). This voltage is defined as an infrequent transient voltage signal that may be applied to the part without damage. Reliability is not degraded from infrequent events but prolonged exposure to the AMV or transient above the AMV may degrade device reliability and even cause device failure. It is usually considered a very short duration event without defining what short means. This nebulous definition is due to the fact

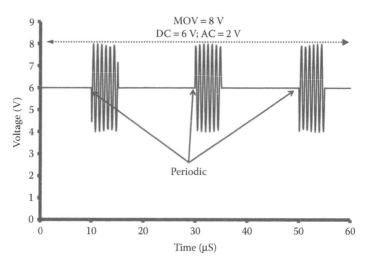

FIGURE 9.1   Illustration of MOV (DC + AC).

that most vendors do not characterize the AMV for each pin. The levels are set based on an understanding of the technology. AMV has to be less than the breakdown of the circuit elements and also lower than the trigger voltage of the ESD protection. Any transient that causes a device to go into breakdown or triggers an ESD clamp has the potential to cause serious damage to the device and must be avoided. The device is expected to function during an AMV transient, but parametric values may not be within specification. Once the transient ends and the voltages return to normal, the part will continue to function without measurable degradation. No voltage experienced in an application should ever exceed the AMV. Beyond AMV, permanent damage or catastrophic failure may occur in the device.

The last key voltage of importance is the device destruction voltage (DDV). The DDV may occur at the voltage that causes breakdown, but it could occur at a higher bias that causes snapback. Figure 9.2 illustrates this with an NDMOS transistor. This is a set of transmission line pulse (TLP) curves at different gate bias (V_{GS}). With $V_{GS} = 0$, the drain voltage increases with each pulse up to the point where breakdown of the device occurs. After breakdown, the voltage and current increase together. The device destruction point is defined as the point at which snapback occurs. The NDMOS transistor shows no damage until this point is reached. With increasing gate bias, the DDV voltage decreases. This behavior is indicative of impact ionization occurring in the drain of the NDMOS device. The impact ionization adds to the channel current. This combination of current aids in turning on the parasitic bipolar formed from the drain, body, and source junctions. NDMOS transistors with very low series resistance may display this behavior. It should be accounted for during the ESD design. The reduced DDV may impact the ESD robustness of the design because the gate voltage may not be zero during an ESD event.

Not all transistors have the same behavior as the NDMOS device discussed earlier. Figure 9.3 shows a regular 5 V NMOS and a ballasted 5 V NMOS device.

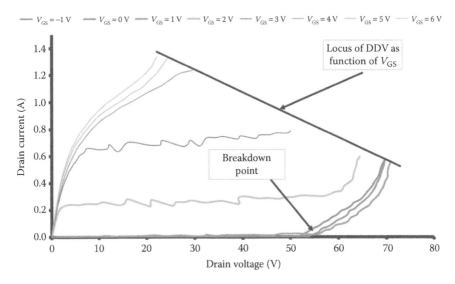

FIGURE 9.2 Safe-operating area (SOA) for NDMOS showing V_{GS}-dependent DDV.

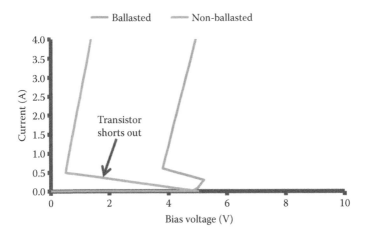

FIGURE 9.3 SOA of ballasted and non-ballasted NMOS.

A ballasted 5 V NMOS has extra resistance placed in series with the drain connection to help limit the current flowing when the parasitic BJT turns on. The ballasting can be internal in the form of a silicide block to remove part of the silicide from the drain/gate edge or externally by placing a resistor in series with each drain finger. The silicide block works best because it removes the silicide metal from the drain/gate edge. This is the location where highest electric field is and also the location where the highest power dissipation will be during breakdown. The silicide metal has a lower melting point than silicon so it is prone to damage at the drain edge. In fact, the regular 5 V NMOS cannot survive snapback because the silicide shorts

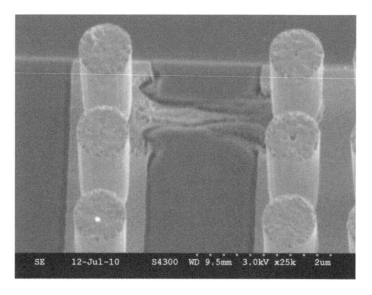

FIGURE 9.4   Drain to source short on a transistor with silicide.

out the drain to body of the transistor and sometimes causes a filament completely across to the source, as shown in Figure 9.4. In the case of the regular 5 V NMOS, the DDV is the breakdown of the transistor. The ballasted NMOS is a special device with respect to ESD. It is used in areas where a regular NMOS would be damaged. As can be seen, the ballasted NMOS supports some ESD current depending on its size and will survive. It can be self-protecting if it is larger than the minimum size needed for the desired current. It is important from an ESD perspective to define the DDV for a circuit element as its smallest voltage magnitude for any bias condition on the device's terminals. Temperature can also play a role because the breakdown and snapback voltage of an element changes with temperature.

The NDMOS shows that DDV is voltage dependent. There are also NMOS devices that are layout dependent. The activation of the parasitic NPN transistor in NDMOS and NMOS transistors is dependent on forward biasing the body to source junction. The channel current flows to the source and does not go into the body to bias the body positive with respect to the source. When the transistor enters saturation, however, impact ionization in the drain can produce a hole current in the body. The location of the body pickup to collect this current influences the body resistance. Large body resistance can make it easier to enter snapback. If the gate voltage is greater than zero, the minimum breakdown may be geometry dependent. The main factor is where the body tie is placed relative to the drain and source. Special test structures are required to measure this behavior. The design rules for the technology define the maximum allowed space from a drain to a body tie. These rules are associated with latch-up but are also important for ESD design. Large body tie spacing makes N-type transistors more susceptible to snapback. If the transistors are not ballasted, they may not survive and would be a source of ESD weakness in the design. As an example, a 0.25 μm 5 V

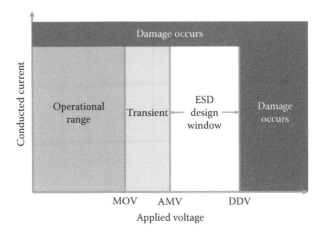

FIGURE 9.5   ESD design window.

transistor may have a DDV of 12 V when the body and source are integrated, but it can drop to 7 V with the body at the maximum of 20 μm away from the drain. The integrated body tie (body and source in same diffusion) results in the highest DDV because the body resistance is minimized.

Taking the MOV, AMV, and DDV together allows one to determine the ESD design window, as illustrated in Figure 9.5. This is the region where the ESD protection has to trigger into its conductive state and provide the level of protection desired. It should be noted that this is for a voltage-level triggered clamp. A transient triggered clamp will have different set of requirements and will be discussed next. The ESD element must not draw significant current at the MOV but must start conducting at voltages above AMV. It must not trigger into its low conductive state at voltages below AMV and also must trigger before DDV and have a dynamic resistance low enough to clamp the voltage below DDV. The ESD window is defined as the voltage difference between AMV and DDV. As ESD protection must trigger at voltages above AMV, it is imperative that no operating voltage or transient voltage exceed AMV, otherwise the ESD clamp could falsely trigger resulting in damage to the circuit.

Transient clamps are a class of clamps designed to respond to the change in voltage as a function of time (dv/dt). There are a number of different types of transient clamps, but most employ a design shown in Figure 9.6 [2,3]. The clamp has three parts: (1) transient detection, (2) driver, and (3) clamping element. Most designs are implemented in a manner where they are not voltage triggered but transient triggered. They do not turn on at specific voltage. The turn-on occurs at a voltage a little above device threshold. Figure 9.6 shows an implementation example of the three blocks. The trigger circuit is a simple R-C network where the capacitor is discharged before the ESD event and is charged during the ESD event. The output of the trigger feeds an inverter that drives the gate of a large NMOS transistor. When the event occurs, the positive rail is charged allowing the gate of the NMOS to be charged through the ON transistor. Current flows through the clamping element until the voltage on the positive rail drops too low to sustain the current flow.

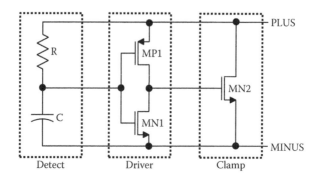

FIGURE 9.6   Basic design of a transient ESD clamp.

The response of the transient clamp to non-ESD and ESD events is equally important. HBM ESD has a rise time as slow as 10 ns, but CDM ESD has a rise time in the 100 ps time frame. It is desired for the clamp to turn on fully and quickly to ensure that these ESD pulses are adequately clamped, but it is not desired for the clamp to turn on during a normal power up or under the influence of noise transients. This places a rise time requirement on the clamps. The author uses a rise time threshold of 1 V/µs as a turn-on threshold for transient clamps. For voltage rise times slower than 1 V/µs, the clamp will not respond and will have a transient current <100 µA. For rise times faster than 1 V/µs, the clamp will fully turn on and provide adequate ESD protection. Each company may have a specific target for their types of products. Another issue with transient clamps is their stability. The single inverter shown in Figure 9.6 is not prone to instability issues, but a three-inverter string can have instability issues [4,5]. Care is needed to ensure that these clamps are stable under all operating conditions.

There are two general types of ESD protections: linear clamps and negative resistance (snapback) clamps. These are illustrated in Figure 9.7. Three examples of a linear clamp are shown. The first is a bigFET transient clamp. There are a number of configurations for this style clamp, but the key clamping mechanism is an ON MOSFET in saturation. Both the gate and the drain are high, and the I–V curve that is mapped out is linear with voltage. The size of the clamp determines the on-resistance. The clamp starts conducting near the threshold voltage of the transistors and dies once the voltage and current reach a critical threshold. Another example of a linear clamp is a diode in both the forward and the reverse direction. In the forward direction at low currents, it looks like a diode, but at high currents the voltage reaches a maximum level and the current continues to increase up to the point of failure. This new conduction mechanism is called *conductivity modulation*. Conductivity modulation occurs when the injected charge alters the bulk diode-free carriers lowering the series resistance. At this point, the resistance in the diode decreases and voltage reaches an isotopic level.

Diodes are most effective in the forward conduction mode but can be used in breakdown for ESD protection. They are not area efficient in the mode of operation. The power dissipation is large during an ESD event because of the larger clamping

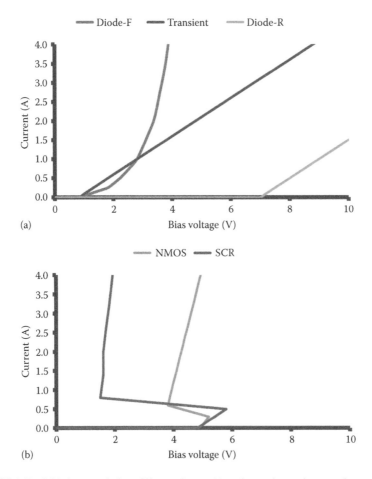

FIGURE 9.7 I–V characteristics of linear clamps (a) and negative resistance clamps (b).

voltage. A breakdown diode-type clamp is illustrated in Figure 9.7a also. In this case, the current is low as the voltage increases to the breakdown point. After the breakdown point, the diode looks like a resistor. The resistance is proportional to the breakdown voltage because of the lower doping needed to achieve the higher breakdown voltages. This drawback can be overcome by adding junctions that inject carriers once breakdown occurs. The injected carriers lower the series resistance but may also lower the voltage across the clamp. These types of clamps are snapback or negative resistance clamps. Examples of these types of clamps are GGNMOS (grounded-gate N-type MOSFET), BJT (bipolar junction transistors), and SCR (silicon-controlled rectifiers). The GGNMOS and SCR are illustrated in Figure 9.7b. Significant work has been done to optimize the snapback clamps to allow them to move toward the ideal clamp [6,7]. The main focus is to improve the holding voltage of these clamps so the change in voltage from the point it triggers to the point it holds is minimized, and the dynamic resistance in the conduction state is near zero.

An ideal voltage trigger clamp will have zero current prior to the trigger voltage and have zero resistance after the trigger voltage. In this mode, it can handle any amount of current by clamping the voltage at the trigger voltage. This should be the goal of each clamp design.

The placement of the trigger voltage within the ESD design window for the different type of clamps is not trivial. The junction breakdowns available in a process will define the trigger voltages for the various clamps. There is limited room to move these trigger voltages. This may limit the AMV and MOV of the technology more than the DDV. Linear clamps because of their ever-increasing voltage with current need to trigger at voltages near AMV to allow the largest ESD window. Their usefulness is then defined by the dynamic resistance, as shown in Figure 9.7a. A lower dynamic resistance will allow a higher current prior to reaching DDV. In contrast, snapback-based clamps tend to be targeted toward the DDV region to allow more room so the holding voltage is greater than MOV. If the holding voltage were below the MOV, significant damage could occur if the clamp was triggered into its holding voltage while the device was powered up. The clamp would not turn off, and large current would flow resulting in damage to the circuit.

9.2.2 Required Circuit Elements

Once the voltage requirements are defined, the next step is to identify the types of circuit elements required to implement the desired function of the product. This task requires a basic understanding of the circuit architecture and how it will operate. At this stage, the detailed circuit schematics are not available, but the designer has a basic understanding of the building blocks that are required and what circuit elements are needed to implement these blocks. Digital and analog designs have different sets of blocks that are needed. As an example, digital designs are made from logic gates, so a library of digital primitives must be available. These include AND, OR, NAND, NOR, INV, and XOR logic gates, just to name a few. Automatic generation of these functional blocks into the smallest physical area is important to a successful design. In contrast, analog blocks are more complex, and their performance is layout dependent. These macros should be manually designed and the layout carefully crafted. Examples of analog blocks are band-gap references, comparators, amplifiers, oscillators, and voltage references. Analog components have the added complexity of requiring matching components for both active and passive devices. The degree of matching can determine the performance of the circuit. Analog circuits may also require trimming to meet the desire performance targets. A trim element allows minor adjustment of a measured parameter such as offset, gain, or oscillation frequency to allow a higher degree of accuracy in the circuit specifications than the process would allow naturally. Table 9.1 shows an example set of categories for a high-voltage power management circuit. The actual table will include all the elements desired for the process with the corresponding specifications required. The detailed list will be different based on the type of circuit needed, but the categories shown in

TABLE 9.1

Example List of Requirements Submitted to Foundry for Quotation Based on 0.13 μm BCD Process

Category	Topics to Check
General process information	Masking layers, photo layers, mask cost, wafer cost (both engineering lots and production), qualification data, years in production, volume of wafers per month, wafer acceptance testing data, roadmap for technology
Transistors	Types and voltage ratings for each transistor available in the process included CMOS, BJT, and high voltage (HV)
Passive devices	Precision and high-sheet rho resistors; high-density capacitors; high- as well as low-voltage capacitors
Memory/trim elements	SRAM block; nonvolatile memory; trim elements
Back end of line (BEOL) features	Number of metal and polysilicon layers; thickness of layers; circuit under pad (CUP) design rules; packaging options
ESD elements	Each voltage node covered; cells included in PDK ready for placement and simulation; ESD verification flow available
Modeling features	Analog-specific model, matching for transistors and passives; voltage and temperature variations included; Monte Carlo modeling for process variability; model versus silicon match
PDK features	Robust PCELL implementation of all elements allowing placement and customization; automated tools for digital blocks as well as memory and other macros
Verification tools	Compatible rule decks for design rule checks (DRC), layout versus schematic (LVS), and electrical rule checks (ERC); verification flow to ensure ESD implemented correctly

Table 9.1 can be used to organize the list for a new development. When developing the list, it is important to prioritize the items based on what would make it easier to design the end product.

The development team need to be realistic in their expectations. The wish list will rarely be fully met. Internally, the development team need to prioritize the list into three categories to help judge the responses from the foundry. These three categories are (1) must have, (2) nice to have, and (3) can live without. The must have category encompasses critical requirements. If this is not available, the project cannot continue. An example of this category is not having the desired high-voltage transistor. The nice to have category is not critical but vital to reduce risk and provide the final design in the fastest time frame. An example of this category would be not meeting the specific on-resistance of the main power transistor. The transistor could be made larger to meet the design targets but would increase the die size. The can live without category falls into the items wished for but not necessary to meet the circuit function or the performance. An example could be that the foundry has a second source for production.

9.2.3   ESD Requirements

Part of the product definition is to define the ESD requirements for the product. This step is often overlooked in the development process but can have significant implications on the viability and cost of the final product if it is not properly specified. ESD requirements come first from the developer and their internal requirements for all products. Each company has a set of minimum ESD requirements that each new product must meet. These can vary from company to company. In general, two ESD models are required to be tested for ESD robustness. These are the human body model (HBM) and the charged device model (CDM) [8–10]. The Industrial Council on ESD Target Levels (Industrial Council) has recommended target levels of robustness (1000 V HBM and 250 V CDM) for both these models, but the industry has not completely accepted these levels for all products [11,12]. One additional component-level ESD model is sometimes required by specific customers. This is called the machine model, but it has been determined to be redundant with the HBM [13]. Without properly specifying the target ESD levels, the final product design may not meet the desired levels.

Some classes of products will require ESD robustness much higher than the target levels defined by the Industrial Council. In addition, some end use products require system-level testing [14]. This type of ESD testing will evaluate products in both a powered and an unpowered state. HBM and CDM testing are performed in an unpowered state. System-level testing can induce voltage transients in the power distribution, which could accidently trigger the ESD protection causing it to turn on. Normal ESD events last for less than 1 μs. In a powered-on state, the clamp could be conducting much longer. Also, ESD has a finite amount of energy available to cause damage. A powered-on system has significantly more energy available for device damage. Figure 9.8 shows an example of a clamp triggered with power applied. The damage is massive and caused the supply to be shorted to ground. The unit was no longer functional. This type of ESD testing places more constraints on the choice of ESD protection and must be considered from the beginning of the project if the project is to be successful.

ESD requirements should be treated just like any other specification. The models that must be passed and the robustness levels for each must be clearly defined upfront so the development team can work to meet these levels. Meeting the ESD needs for the product may require a separate development effort in parallel with the product development if there are special ESD requirements. Some circuits by the nature of their application will require special ESD. As an example, it is common for interface products to require very high ESD robustness because they must provide a communication path between one system and another system. Examples of these are RS-485, RS-232, and SERDES circuits as well as Ethernet and surveillance video and audio. Existing products in the marketplace already have very high ESD robustness. This poses a barrier for new products because they must achieve the high ESD robustness to compete with the existing products.

The type of circuit may also limit the choice of an ESD protection strategy. High impedance pins such as the inputs to operational amplifiers (op-amps) or comparators are sensitive to leakage. Leakage above 100 nA may be too high for them. Typical ESD elements may have leakage currents too high to meet these

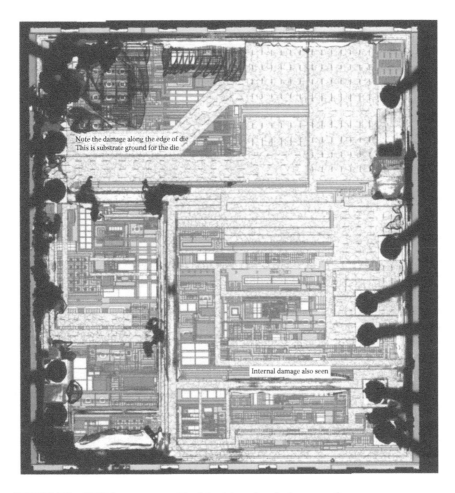

FIGURE 9.8 EOS damage caused by false triggering during operation.

leakage requirements due to their physical size. A smaller size would not meet the ESD requirements. In these cases, a new ESD structure is needed that has both high robustness and small junction area. An SCR is one of the most area-efficient clamps available to the ESD designer and could be used for this type of pin.

High-speed operation can also pose challenges for ESD protection. The parasitic capacitance added by the ESD may distort signal fidelity to the point the circuit no longer functions. Serial interface speeds have increased making this aspect even more important. The absolute capacitance is not the only issue. Signal fidelity is also affected by how the capacitance changes as a function of voltage. It is desired for the capacitance to be small and constant with respect to voltage. This may be hard to achieve without combining multiple elements in parallel to balance the capacitance as the voltage changes. Achieving the desired signal fidelity and ESD robustness is a balancing act pitting performance with ESD robustness. In many cases, performance wins and ESD robustness are degraded to achieve the desired performance.

Open drain pins are another pin type that requires special attention. These pins do not have a path to the positive supply. Some inputs have a similar requirement of no conduction path to a positive supply. One such type is a voltage-tolerant pin. This class of pin may allow 5 V on the pin but be powered from a 3.3 V supply. A diode connection from the 5 V pin to the 3.3 V supply would draw significant current. These types of pins cannot use a dual-diode ESD protection because of the diode to the positive supply. These pins rely on a clamp from the pin to ground. It is important that the clamp turn on prior to the voltage reaching the DDV of any circuit element tied to the pin. This type of protection may not be readily available in the process and would require development.

9.3 PROCESS SELECTION

In the early stages of a product development, there is a general idea about what type of process is needed, but after defining all the requirements the act of selecting the best process to use is difficult. The decision is based on getting the most benefit from the process at the lowest cost. The analysis of these two factors (benefit and cost) is not as easy to quantify as one might think. The first topic of discussion is the technology node selection. The technology node defines the level of integration and usually the smallest feature size in the process. One might think that the smallest node would yield the most performance. It may yield some of the fastest transistors, but it carries some drawbacks as well. The primary drawback is cost. The most advanced process costs the most in terms of mask set cost (nonrecurring cost) as well as wafer cost (recurring cost). It is possible that the cost associated with the most advanced technology may be prohibitive for some projects. There are also intangible costs associated with a process. These costs include the ease of working with the foundry, the amount of automation and flexibility built into the product design kit (PDK), the stability of the foundry, loading in the foundry, and the ability to have a second-source wafer fabrication facility separated from the primary site as a catastrophe mitigation strategy. All these costs play a role in the process selection because they play a role in the ease of designing products in the process as well as the ability to get silicon out of the foundry in a timely manner with the necessary volume to meet your customer's demands.

Selecting the optimum technology node starts with understanding the type of circuit and what circuit elements are needed.

Digital circuits benefit the most from small geometries and large numbers of interconnect layers. They can be packed very densely occupying the smallest space. Analog circuits are less dense and need special layout techniques to achieve optimum performance. They also need noise isolation between blocks and signal lines making their overall density even less. Last, high-voltage elements form another class of circuit. These circuit elements are the least dense and take up the most area. The operating voltage defines the area needed to support the breakdown of the device. When deciding on a technology, it is important to consider how much of each circuit type will be needed in the product. If the circuit is primarily digital with just a little analog, a more advanced technology with smaller dimensions would allow the smallest die size and the small size could justify the larger wafer cost. On the other hand, if most of the product is a high-voltage device and some analog circuitry, there will be little benefit going with a very advanced technology. It is important to understand the mix of these elements in the planned products and to select the technology node that provides the smallest cost per die.

The cost per die is calculated from the die size, wafer size, and cost per wafer. The die size and the wafer size define the number of die per wafer. The wafer cost divided by the number of die per wafer yields the die cost. As the die size increases, the number of die per wafer decreases and the cost per die increases. The wafer cost is technology dependent so as the technology scales to smaller geometries, the wafer cost increases. Wafer cost is also an increasing function of the number of features in the technology such as number of metal layers and number of circuit elements. Die size may also shrink with more metal layers because the metal connections between circuit elements can be made more efficient taking up less space.

The size of the ESD protection directly adds to the die size and therefore the die cost. The ESD area is an overhead function because it does not add functionality or performance to the product. It is there for protection only. It is always desirable to minimize the area allocated to ESD and use just the size necessary to achieve the desired ESD performance. Excess in ESD performance wastes area and increases die cost. Older products could afford a conservative approach for ESD, but it is more costly in advanced technology. The extra space occupied by more robust ESD could be used for extra functionality at the same die size, or the die size could be reduced resulting in a lower cost product. It should be noted that ESD robustness also decreases with technology node so more advanced technology will be more difficult to provide adequate ESD protection [15].

The selection of the process is a combined effort from the purchasing, technology development (TD), design, and marketing. The technical part of the selection process is driven by the TD organization and purchasing drives the price negotiations. This process starts with a statement of work developed jointly by TD and design, detailing what is needed in the process. A request for proposal (ROP) is sent out to multiple foundries looking for bids to supply the desired technology. Some organizations will NO BID the process. They are not interested in supplying the desired technology. Others will supply a bid, but it is rare that any single organization will provide everything that is desired. The team should expect this going into the process and have internally prioritized list of must have, nice to have, and can live without. These categories can help weed down the list of potential candidates to a small number making it easier to review. The desire is to get as close to the desire list as possible and identify the organizations that are most likely to develop the remaining elements or have something close enough to be usable for the final product.

9.4   ESD ASSESSMENT

The ESD assessment reviews the readiness of the process from an ESD design perspective. It is best to do this assessment as a part of the process selection task. Doing the ESD assessment at the process selection stage allows ESD to be a part of the selection criteria. It is common that reviewers inexperienced in ESD will see the term *ESD* in the documentation and maybe even some elements in the PDK labeled as ESD protection and assume it meets all their needs. A detailed ESD assessment is the only way to really understand what and where the deficiencies are for the specific types of circuits desired. The importance of doing the ESD assessment at the process selection stage may not bubble up to the necessary level of importance. If it is not a part of process selection, then an ESD assessment must be performed soon after the process is selected. The longer the ESD assessment is delayed, the more difficult it will be to recover should deficiencies be determined. Doing the ESD assessment early may also save significant time if the chosen process is significantly lacking with respect to ESD. A new process may need to be selected, or significant ESD development must be initiated.

The ESD assessment reviews all aspects of the technology with respect to ESD and looks for deficiencies that will make ESD protection implementation difficult or impossible. The assessment must cover the types of ESD protection available in the process as well as the robustness of each circuit element and interconnecting layers in the process. It also must assess the ease of implementation and the ESD checking strategy to ensure that ESD is implemented correctly. Table 9.2 is a checklist of ESD-related things needed on a typical power management process. This is only a sample listing and would be customized for each type of product and technology.

TABLE 9.2
ESD-Related Foundry Checklist for Power Management Process

Category	Topics to Check
ESD protection diodes	PCELL based; scalable; N+/PWELL and P+/NWELL; triple-well versions recommended; isolation technique
Transient clamp designs	PCELL based; scalable in size for different ESD levels; PMOS and/or NMOS clamping element; characterized for leakage
Ballasted transistors	PCELL based; used in output transistors for self-protection; ESD implant to reduce trigger voltage for stand-alone clamp; optimized ballasting and body tie spacing to minimize size
ESD clamps	PCELL based; need protection for each voltage node; characterization report showing ESD robustness and leakage
Floating clamps	PCELL based; ESD protection for high-side driver (floats above substrate); usually low-voltage ESD but can float to HV
Ground protection	Ground-to-ground protection from power to analog domain (noise isolation, 3–4 V of noise)
Circuit element assessment	TLP characterization of ESD robustness for each circuit element; used to determine sensitivity for each element
BEOL interconnect robustness	TLP characterization of metal, polysilicon, and any other conductive layer in the circuit that could carry ESD current
ESD design rules	Detail-specific ESD rules that need to be followed for the process and how these rules are checked for a physical design
Modeling data	ESD clamp model capability; can ESD simulations be performed or only parasitic effect on circuit

The ESD assessment needs to be performed with the end circuits and ESD requirements in mind. This is especially true if there is a list of special circuit needs or special ESD requirements. This is not a generic ESD assessment of the selected process but one specifically targeted at the requirements previously defined. Pins that require low leakage and/or low capacitance may not be available. If these pins are integral to a successful development and they do not exist at the beginning of the project, it adds a lot of risk to the schedule because they would need to be developed in parallel with the product development. It is important to identify these risk areas early in the development where there is time to react and implement alternative strategies to mitigate these risks.

Finding the information needed to make the ESD assessment may not be easy. It is rare that a foundry has everything in one place. This reality could change in the future because of the work done by the ESD Association (ESDA) [16]. In 2014, the ESDA published ESD TR22.0-01-14, which is the first document to guide process suppliers with the type of information needed by process users with respect to ESD development and implementation. As foundries start providing the information in this technical report, it will be easier to assess a foundry's ESD capability and also to compare its robustness to other foundries.

The primary source of technical information for the process is found in the documentation for the design kit. This documentation should also have information about ESD design and implementation. The design kit documentation may have a process flow or description showing the sequence of steps needed to construct the wafer. This document is useful to understand the construction of the transistors and clamps and may be beneficial if new or modified designs are needed. It will also have documents overviewing the design techniques used for this process called a Designer's Guide. This document has electrical specification for the circuit components and also documentation about how to use the circuit components. The Designer's Guide may also have detailed information about the ESD clamps available in the process and how to properly apply these clamps to new circuits. One key document that is needed with any process is the design rule document. The design rules govern the size, spacing, and placement of geometries on a wafer. It provides the details about how the different layers in the process form the circuit elements and how they are placed and connected to each other to make a working product. More advanced technologies have two options available for the spacing and overlaps of components. The first is the minimum spacing allowed. These options are based on the photolithography tools used to define the layers in the process. There are a more relaxed set of rules termed *design for manufacturing* (DFM) rules. The DFM rules do not push the process as hard with respect to tolerance so the process may yield better, and there will be less variability in the process. It is recommended to use the DFM design rules for ESD elements.

There are a number of other documents that are useful for the ESD designer. Parasitic extraction reports can provide details about the sheet resistance of the metal and polysilicon layers used in the process. This is important when calculating the series resistance between pads and clamps. Wafer acceptance testing reports contain the electrical data that monitors the process parameters for each diffusion lot. This report should include the measured breakdowns of each circuit element in the process. The breakdowns of each transistor and junction are a good starting point to determine the trigger voltage of the clamps for each voltage node in the process. If the ESD protection can clamp the voltage below the breakdown of each element, the likelihood of circuit damage from ESD is very small. Finally, the reliability qualification report will include the gate-oxide integrity (GOI) plots showing the distribution of gate-oxide rupture voltage. This data can provide insight into the robustness of the gate oxide under ESD stress. This data is taken with a slow voltage ramp. ESD is a very fast event, with HBM ESD being the slowest. The rise time of HBM ESD is <10 ns and the duration is ~750 ns. Under these ESD conditions, the failure voltage for gate-oxide rupture increases about 1.5 times the value reported under a slow ramp [17]. CDM is even faster with rise times in the 100 ps time frame and durations in the 1–5 ns duration. The rupture voltage under CDM events will be even higher [18]. It is important to review the gate-oxide rupture voltage with respect to the junction breakdown voltage. For advanced technologies, the gate dielectric may be so thin that it breaks down at voltages lower than the junction breakdown. In this case, junction breakdown devices cannot adequately protect the gate oxide, and a different approach is needed. The most effective method of protecting the gate oxide is a dual-diode secondary clamp, as shown in Figure 9.9. In this clamp, a voltage dissipation resistor is placed prior to the dual-diode clamp to drop the voltage. The size of the resistor is inversely proportional to the diode size. Small diodes require a larger resistor to limit the current. Here the local gate-oxide voltage is clamped to within a diode drop of the local supply voltage. It should be noted that this method is useful for both I/O pins on the part and internal cross-domain cells. Also, the use of isolated NMOS transistors can improve the CDM robustness of transistors significantly by shielding the gate oxide from the direct application of the charging voltage, as shown in Figure 9.10 [19]. In Figure 9.10a, the substrate forms one plate of the gate-oxide capacitor, allowing full potential across the gate oxide during a CDM event. In the case of Figure 9.10b, the isolation forms an additional two-series capacitors, which can drop the voltage applied to the gate oxide and also provide additional discharge paths.

9.5 FILLING THE GAPS

After assessing the technology for ESD robustness, there will be gaps in the knowledge or protection that require filling. The gaps may come in the form of missing protection for certain voltage nodes, ESD-specific PCELLS in PDK, or undocumented circuit robustness. Documenting these deficiencies is a very important step for the development team. The list documents what is needed and can form a prioritized risk mitigation plan. Also, if the list is extensive, it may prompt a change in the decision on which process to use for the development.

Once a prioritized list is developed, the team must decide what task is required to address each deficiency. There are a number of options available to the team, and each requires a balance of development cost, time to market, and resources. The resources needed to develop something new may be extensive as well as time prohibitive. In some cases, the foundry partner may be willing to meet the

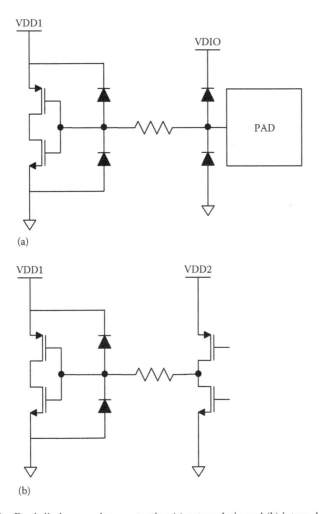

(a)

(b)

FIGURE 9.9 Dual-diode secondary protection (a) external pin and (b) internal net.

deficiency. This may be the lowest financial cost option for the team, but it comes with an intangible price. The foundry partner is able to offer this enhancement to a competitor as well.

The ESD needs of most pins on a new part may be satisfied by a general-purpose input/output (GPIO) library. These are standard-size cells with a fixed height and varying width based on their function. Figure 9.11 shows an example of these types of cells. These cells are provided by the foundry or a third-party I/O vendor. They are very good at protecting digital pins that use normal I/O and core voltages but may have deficiencies when it comes to high-speed pins, voltage-tolerant pins, analog pins, or high-voltage pins. In general, GPIO cells meet ~80%–90% of the ESD needs for a product. The remaining 10%–20% must be filled by other means. One of the first options is to use a custom ESD vendor for these special pins. They may

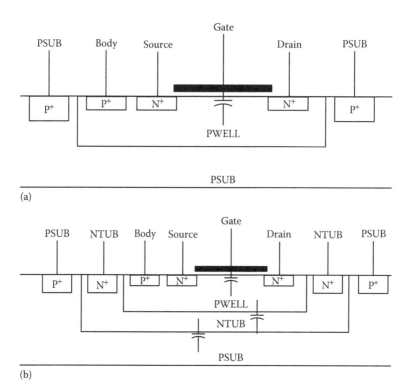

FIGURE 9.10   Non-isolated (a) and isolated (b) transistor for CDM protection.

have off-the-shelf designs to meet the special ESD needs such as low capacitance, low leakage, and pins with high ESD robustness. There are a number of IP vendors for ESD who have proven solutions at key foundry suppliers. The benefit is time to market. They may have a working cell that can be incorporated immediately into the project flow. The drawback with these types of vendor is that they require special licensing and royalty payments to use their IP. This adds both nonrecurring and recurring cost and will lower the overall profitability of the product.

Another option is to develop the ESD solution internally. This is only an option if the development team have the proper support organization to develop the ESD, model it, and then implement it in the PDK. Another aspect of this approach is the time required to accomplish these tasks. The ESD design team will need to develop and evaluate multiple test chips to learn the process and optimize the clamps. It is best to plan on at least two test chips to have clamps validated in silicon. From a time perspective, this could easily take 9–12 months of time depending on the cycle time through wafer fabrication. The product team may not have this amount of time to get the product to market. The advantage of the internally developed option is the ESD IP is kept internal to the company and is not shared with the foundry or any third party. This may be the best option if unique ESD is required. The drawback is the time and cost to develop this type of protection.

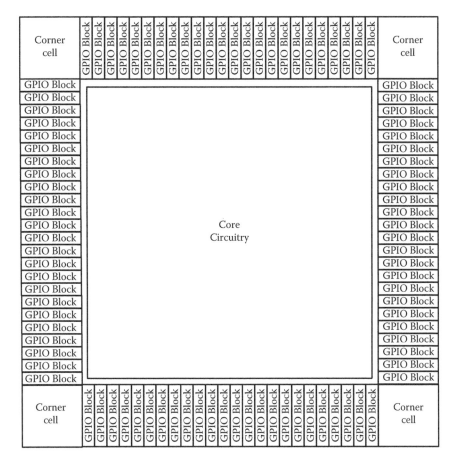

FIGURE 9.11 Example of a GPIO cell library.

An early ESD assessment of the process can identify the ESD needs and include them in the business development plan. Identifying these needs early allows for a parallel ESD and product development. It should be noted that the perfect solution for ESD may not exist. In general, the best option is to use a hybrid approach where most of the pins are protected from a GPIO library and specific pins are protected by internally developed ESD or from a special ESD vendor. This combination provides the lowest cost and best time to market.

9.6 CONCLUSIONS

ESD design in a foundry process can be challenging. The development is fraught with unknown, incomplete coverage of voltage nodes and compromises between performance and ESD robustness. The development team should not assume that the foundry has adequate ESD just by seeing the acronym "ESD" in the documentation. An extensive evaluation of the foundry's ESD capability is needed. This review is targeted at the needs for the specific design being developed. If the ESD capability is lacking, the team may need to choose a different foundry to supply the process. If the decision is to stay with the foundry, a mitigation plan needs to be developed and implemented to reduce the risk associated with product release. ESD development can be successfully done in a foundry, but the development team need to enter the relationship fully understanding the technical and schedule risks associated with ESD at the foundry of choice.

REFERENCES

1. Transistor count, 2015. Retrieved from http://en.wikipedia.org/wiki/Transistor_count.
2. E.R. Worley et al., "Sub-micron Chip ESD Protection Schemes which Avoid Avalanching Junctions," in *Proceedings of the EOS/ESD Symposium*, Vol. EOS-17, pp. 13–20, 1995.
3. C.A. Torres et al., "Modular, Portable, and Easily Simulated ESD Protection Networks for Advanced CMOS Technologies," in *Proceedings of the EOS/ESD Symposium*, Vol. EOS-23, pp. 82–95, 2001.
4. B.L. Hunter, B.K. Butka, "Damped Transient Power Clamps for Improved ESD Protection of CMOS," *Microelectronics Reliability*, Vol. 46, pp. 77–85, January 2006.
5. H. Sarbishaei, "Electrostatic Discharge Protection Circuit for High-Speed Mixed-Signal Circuits," PhD dissertation, Electrical and Computer Engineering, University of Waterloo, Ontario, Canada, 2007.
6. J.A. Salcedo, J.J. Liou, J.C. Bernier, "On the Design of Tunable High-Holding-Voltage LVTSCR-Based Cells for On-Chip ESD Protection," in *Proceeding of the 7th International Conference on Solid-State and Integrated Circuits*, Vol. 2, pp. 798–803, 2004.
7. Z. Liu, J. Vinson, L. Lou, J.J. Liou, "An Improved Bidirectional SCR Structure for Low-Triggering ESD Protection Applications," *Electron Device Letters*, Vol. 29, No. 4, pp. 360–362, 2008.
8. "Electrostatic Discharge Sensitivity Testing Human Body Model (HBM)—Component Level," ANSI/ESDA/JEDEC JS-001-2014, August 28, 2014.
9. "Electrostatic Discharge Sensitivity Testing—Charged Device Model (CDM)—Component Level," ANSI/ESD S5.3.1-2009, December 4, 2009.
10. "Field-Induced Charged-Device Model Test Method for Electrostatic-Discharge-Withstand Thresholds of Microelectronic Components," JESD22-C101F, October 2013.
11. "Recommended ESD Target Levels for HBM/MM Qualification," JEP155A.01, March 2012.
12. "Recommended ESD-CDM Target Levels," JEP157, October 2009.
13. "Discontinuing Use of the Machine Model for Device ESD Qualification," JEP172, July 2014.
14. "Electromagnetic Compatibility (EMC)—Part 4-2: Testing and Measurement Techniques—Electrostatic Discharge Immunity Test," IEC 61000-4-2 Ed. 2.0 b:2008, December 2008.
15. *Electrostatic Discharge (ESD) Technology Roadmap*, March 2013, Retrieved from https://www.esda.org/assets/Uploads/docs/2013ElectrostaticDischargeRoadmap.pdf.
16. "Relevant ESD Foundry Parameters for Seamless ESD Design and Verification Flow," ESD TR22.0-01-14, 2014.
17. A. Ille et al., "Ultra-thin Gate oxide Reliability in the ESD Time Domain," in *EOS/ESD Symposium*, pp. 285–294, 2006.
18. S. Malobabic et al., "Gate Oxide Evaluation under Very Fast Transmission Line Pulse (VFTLP) CDM-Type Stress," in *Proceedings of the 7th ICCDCS*, pp. 1–8, 2008.
19. J.H. Lee et al., "The Study of Sensitive Circuit and Layout for CDM Improvement," in *IEEE Proceedings of the 16th IPFA*, pp. 228–232, July 2009.

10 Compact Modeling of Semiconductor Devices for Electrostatic Discharge Protection Applications

Zhenghao Gan and Waisum Wong

CONTENTS

10.1 INTRODUCTION

With the continuous scaling of semiconductor technology, electrostatic discharge (ESD) protection becomes more challenging. The commonly used devices for ESD protection are resistor, diode (STI diode and gated diode), bipolar junction transistor (BJT), gate-grounded MOS (GGMOS), and silicon-controlled rectifier (SCR), with some modifications of the standard device layout and design rule, to address the ESD requirements. Figure 10.1 gives a schematic of basic on-chip ESD protection concept. The ESD clamp can be diode, BJT, GGMOS, or SCR. Resistor (R_{ESD}) is basically located between the primary and secondary ESD protection to shunt large current.

Accurate ESD device models are desirable for designers to simulate the effect of the ESD devices on the overall circuit performance and to perform co-design of the ESD and functional circuit. This becomes particularly important with the continuous reduction of process margin. The accurate ESD device models also make it possible for the circuit designers to predict the protection level of a given chip design prior to fabrication and test.

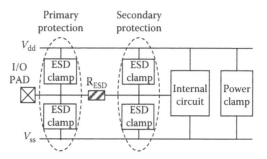

FIGURE 10.1   Schematic of basic on-chip ESD protection concept.

Of course, the process design kit (PDK) provided by the foundry includes the base models for resistor, diode, BJT, and MOS devices under normal operation. However, in case that these devices operated under ESD high-current injection, their current–voltage (I–V) behaviors are much different from and more complicated than those under normal operation. Furthermore, the devices used in protection circuits may be processed and layout tuned for ESD reliability and, thus, not well modeled by parameters extracted for devices used under normal operation. As a result, compact modeling of semiconductor devices for ESD protection applications is necessary, which is the topic of this chapter.

There are many efforts made in both academic and industry to make circuit-level simulation models for devices working under ESD conditions available. In literatures, several research groups have proposed models of the most commonly used devices in ESD protection circuits, namely, diodes [1–4], metal–oxide–semiconductor field-effect transistors (MOSFETs) [4–10], and resistors [11].

In this chapter, compact models for resistors (diffusion resistor and metal resistor), diodes (STI diode and gated diode), GGMOS, and vertical BJT under ESD high-current injection are detailed. With these components' models at hand, Simulation Program with Integrated Circuit Emphasis (SPICE) simulation to evaluate a given circuit performance under ESD stressing is easy to perform prior to fabrication and test.

10.2   RESISTOR MODEL

10.2.1   Model for Diffusion/Well Resistor under ESD Stressing

Diffusion and well resistors are widely used in ESD protection circuits to shunt large current between the primary and secondary ESD protection (i.e., R_{ESD} in Figure 10.1). Typical value of R_{ESD} is 100–300 Ω [6]. For high-speed I/O, R_{ESD} should not be large enough to cause an appreciable RC delay. In general, N-type resistors are preferred for use with ESD circuits in processes that use P-substrate and P-type resistors with N-substrate [6]. There is basically no difference between the ESD resistor and the regular resistor. A typical top and cross-sectional view of a diffusion resistor (with and without silicide) is shown in Figure 10.2.

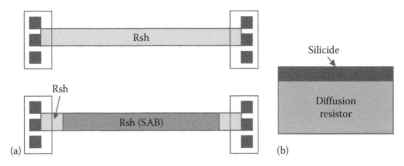

FIGURE 10.2 A typical (a) top and (b) cross-sectional view of a diffusion resistor (with and without silicide).

Using N-type resistor as an example, its current density (J) at low injection current (ohmic) is expressed by neglecting the minority hole current as follows:

$$J \approx J_n = q\mu_n n_c E = q n_c \upsilon_d \qquad (10.1)$$

where:
q is the electron charge
μ_n is the electron mobility
n_c is the doping concentration
E is the electric field
υ_d is the drift velocity

Figure 10.3 shows typical I–V characteristics of N-diffusion resistor under ESD pulse high-current injection. When the current is low, the resistor presents linear

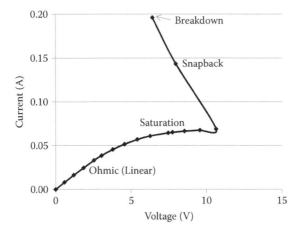

FIGURE 10.3 Typical I–V characteristics of N-diffusion resistor under ESD pulse high-current injection. The resistor is 4 μm wide and 4 μm long.

current versus voltage dependence, that is, ohmic behavior. As the I–V increase, the current becomes far below the linear-dependent value and turns to be saturated because of the carrier drift velocity saturation. In general, the electron drift velocity saturates at $\upsilon_{sat} = 10^7$ cm/s at $E = 10^4$ V/cm. Further increase in voltage will not increase the J value as

$$J = J_{sat} = qn_c\upsilon_{sat} \qquad (10.2)$$

However, when the voltage increases even further into deep saturation region, a snapback behavior is observed. One explanation is that when the electric field in the resistor is large enough and reaches the impact ionization threshold, electron–hole pairs (EHPs) are generated. When the generated hole current becomes large enough to contribute to the total current, the supported voltage decreases, and a negative resistance or snapback characteristic appears. The amount of impact ionization required to cause snapback depends on the doping concentration, which would influence the maximum voltage across the resistor before snapback occurs. Another explanation is that the snapback is attributed to the self-heating due to large current injection [12].

The modified high-field resistance accounting for conductivity modulation associated with high injection of minority carriers or mobility degradation because of high fields under transmission line pulse (TLP) measurement can then be written as [6]:

$$R_{tlp} = R_0\left[1+\left(\frac{V_r}{V_{sat}}\right)^n\right]^{\left[1+(1/n)\right]} \qquad (10.3)$$

where:
R_0 is the low-field resistance (a fitting parameter)
V_r is the voltage across the resistor
n is a fixed constant (~1)
V_{sat} is a fitting parameter

10.2.2 MODEL FOR METAL INTERCONNECT UNDER ESD STRESSING

High joule-heating during ESD transient can cause failure in metal interconnects. The metal interconnect becomes a factor limiting the ESD capability for other components (diode, GGMOS, etc.). This is especially true for advanced technology node in which the metal is thinner and current density in the metal will be higher for the same current level going through the metal. The low-k or ultra-low-k materials used for dielectric insulation lead to slower heat dissipation, which also cause metal failure.

As mentioned earlier, when the metal interconnect is subject to ESD pulse stressing, its current density will increase, which further increases the line temperature

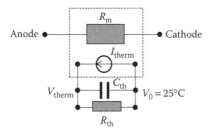

FIGURE 10.4   Sub-circuit model for metal interconnect under both high-current TLP.

due to heat generation by the power ($P = I^2 \times R$). This is so-called self-heating effect (SHE). The temperature increase of the metal line will further lead to a higher resistance (R), which results in an even higher heat generation, higher temperature, and higher resistance. With this positive feedback, the metal interconnect approaches its melting point at certain current density, leading to the final breakdown if the ESD pulse stressing continues. This process is thus destructive [13,14]. The SHE process is complicated as the metal line is embedded in a dielectric environment that retards the heat dissipation generated in the metal line. This means that the situation is much dynamic during the ESD pulse stressing.

Thus, it is crucial for supporting ESD designers to develop a scalable compact model able to take SHE into account and include a failure criterion to predict the transient and quasi-static behavior of an interconnection as well as its failure [15].

Figure 10.4 shows the sub-circuit model for metal interconnect under high-current TLP. The metal interconnect itself is represented by the component R_m. In this figure, a thermal equivalent circuit is constructed for the determination of resistor temperature rise due to SHE. Each of the electrical quantities shown in the figure has a thermal equivalent. I_{therm}, V_{therm}, V_0, R_{th}, and C_{th} correspond to power ($P = V_{Rm} * V_{Rm}/R$), the device temperature due to SHE (T), chip reference temperature (e.g., 25°C for room temperature), thermal resistance, and thermal capacitance, respectively. V_{Rm} is the voltage across the resistor R_m. The equation for this circuit can be expressed as

$$I_{therm} = C_{th}\frac{dV_{therm}}{dt} + \frac{V_{therm} - V_0}{R_{th}} \tag{10.4}$$

where:

$$I_{therm} = \frac{V_{Rm}^2}{R_m} \tag{10.5}$$

By solving Equation 10.4, the temperature increase (ΔT) in the metal interconnect due to SHE can be expressed as:

$$\Delta T = V_{therm} - V_0 = I_{therm} \cdot R_{th}\left[1-\exp\left(-\frac{t}{R_{th}\cdot C_{th}}\right)\right] = \frac{V_{Rm}^2}{R_m}\cdot R_{th}\left[1-\exp\left(-\frac{t}{R_{th}\cdot C_{th}}\right)\right] \quad (10.6)$$

In a more conventional way, the metal resistance R_m can be expressed in terms of its ΔT, TCR (temperature coefficient of resistance), and its resistance at room temperature under low field (R_{0m}):

$$R_m = R_{0m}\cdot\left(1+TCR\cdot\Delta T\right) \quad (10.7)$$

When inputting Equation 10.6 into the above equation, we get,

$$R_m = R_{0m}\cdot\left\{1+TCR\cdot\frac{V_{Rm}^2}{R_m}\cdot R_{th}\left[1-\exp\left(-\frac{t}{R_{th}\cdot C_{th}}\right)\right]\right\} \quad (10.8)$$

Equation 10.8 shows the metal interconnect resistance after an ESD pulse stressing of time ($t = 100$ ns in a general case). The $R_m(t)$ depends on the voltage applied, TCR, R_{th}, and C_{th}.

Figure 10.5 shows the comparison of the silicon data (the stars) and the model fitting (the curve) based on the sub-circuit given in Figure 10.4, focusing on high-current injection (TLP). The model can match the silicon data very well. It is noted that there is a bit deviation of the model from the silicon data when the voltage >10 V. This is caused by an underestimated TCR at high temperature as explained later. As shown in Table 10.1, it is seen that ΔT can be as high as 470°C, meaning its temperature is ~500°C, which is approaching the aluminum melting temperature of 660°C. This is the reason leading to the breakdown of the metal line when its voltage applied is ~13 V. In the model, we use TCR obtained up to 200°C. It is clear that

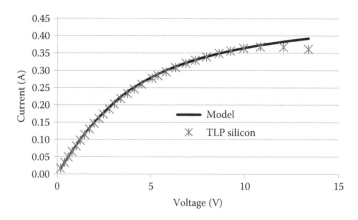

FIGURE 10.5 Comparison of the silicon data (the stars) and the model fitting (the curve) based on the sub-circuit given in Figure 10.4, focusing on high-current injection (TLP) for M1 with $L = 100$ μm.

TABLE 10.1

The Voltage across the Metal Interconnect under Study (Figure 10.5) and the Corresponding Temperature Increase (ΔT) and Its Resistance Change

V	ΔT (°C)	R (Normalized)
0.19	0.3	1.0
1.49	15.6	1.1
2.45	39.1	1.2
3.38	67.8	1.3
4.52	107.9	1.4
5.38	140.3	1.5
6.30	176.6	1.7
7.36	219.4	1.9
8.61	270.7	2.1
9.92	325.7	2.3
10.80	362.9	2.4
12.04	415.4	2.6
13.35	471.8	2.8

Note: The data is from the model.

the TCR should be temperature dependent and much higher when it is approaching the melting temperature. Therefore, the model deviates from the silicon data when voltage >10 V.

10.3 DIODE MODEL

Diodes are widely used for ESD protection because they are robust and easily implemented. STI diode and gated diode are two typical diodes, as schematically shown in Figure 10.6. The dual-diode circuit (Figure 10.7) is frequently used at I/O pads. STI diode is more commonly used for ESD protection. However, with the continuous scaling

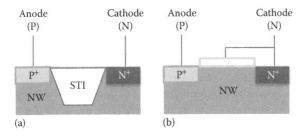

FIGURE 10.6 Schematic cross section of P+/NW diode: (a) STI diode and (b) gated diode.

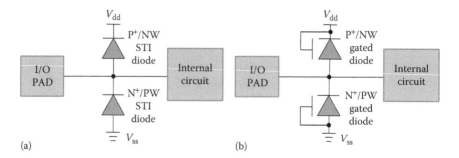

FIGURE 10.7 Schematic of dual diodes in a circuit used for ESD protection: (a) STI diode and (b) gated diode.

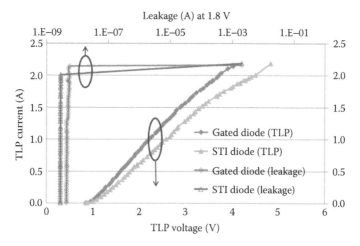

FIGURE 10.8 Comparison of the TLP I–V characteristics of gated diode and STI diode with the same finger width and finger number.

of semiconductor technology, gated diode has been recognized as one of the options because of the following characteristics [16] compared with STI diode: higher TLP I_{t2} due to lower series resistance, better Q-factor for RF application, and more robust CDM performance. As shown in Figure 10.8, the STI diode has a higher on-resistance but a lower breakdown current (I_{t2}) than the gated diode. However, gated diode has a drawback that its leakage is relatively higher than STI diode under operation.

Figure 10.9 shows the schematic top view of the STI ESD diode structure under study including multi-finger metal interconnection and the multi-finger P+/N-well diode. The layout geometrical parameters are LR (length of active area [AA] with P+ diffusion), WR (width of AA with P+ diffusion), NF (finger number of diode), WM (width of M2/M3), and NM (finger number of M2/M3).

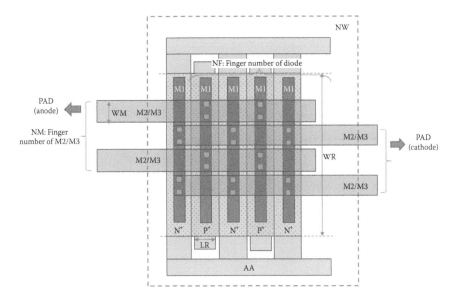

FIGURE 10.9 Schematic top view of the ESD diode structure including PAD, multi-finger metal interconnection, and the multi-finger P⁺/NW diode.

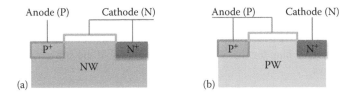

FIGURE 10.10 Schematic cross section of the gated diodes under study where the gate is connected to the well: (a) P⁺/NW and (b) N⁺/PW.

ESD gated diodes with multi-fingered structures were fabricated using CMOS technology. Figure 10.10 shows the schematic cross section of the P⁺/NW and N⁺/PW gated diode structures under study where the gate is connected to the well. Figure 10.11 gives the schematic of the top view of the gated diode with multi-fingers. Layout splits with finger width (W), finger number (NF), and gate length (L) are studied and covered in the compact model.

Measurements were performed using two different tools. In the conventional DC bias (low-current regime <0.1 A), an HP4156 parameter analyzer was used. In the high-current regime, TLP measurement was done using Barth 4002 test system with 100 ns pulse duration.

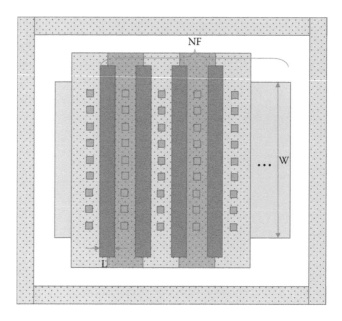

FIGURE 10.11   Schematic of the top view of the gated diode with multi-fingers: W, finger width; NF, finger number; L, gate length.

10.3.1   STI Diode Model

Figure 10.12 shows the DC equivalent circuit SPICE model covering I–V characteristics under low DC bias and ESD TLP (high-current injection) and I–V characteristics under reverse bias for the STI diode. The details of the sub-circuit model are given as follows:

- Anode is connected to P, and cathode is connected to N.
- *Diode*: the conventional diode model.
- R_{tlp}: voltage-controlled substrate resistance to accurately model the substrate conductivity modulation associated with high injection of carriers due to mobility saturation during high-current TLP, as described by Equation 10.3.
- R_m: voltage-controlled metal resistance to accurately model the metal resistance increase due to SHE with high-current injection, as detailed in Equation 10.8.

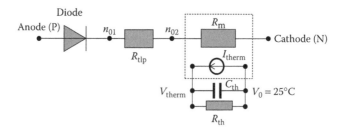

FIGURE 10.12   The DC equivalent circuit SPICE model under ESD TLP for the STI diode.

Figure 10.13 compares the model fitting with (1) conventional diode, (2) additional R_{tlp}, and (3) additional R_{tlp} and R_m to the silicon data under ESD high-current injection. The measurement data are plotted with dots while the models are plotted with curves. As shown in Figure 10.13a, there is large discrepancy at high current where the model is fitted based on the conventional diode. Further, when adding R_{tlp} considering substrate resistance modulation under large current, the fitting of the I–V between 2 V and 4 V is improved, as shown in Figure 10.13b. However, when the voltage is larger than 4 V, it looks that the measured current under TLP measurement is saturated, which is much smaller than the model. This behavior is similar to the metal interconnect under ESD pulse stressing, as shown in Figure 10.5. Taking this in mind, as shown in Figure 10.13c, R_m considering metal SHE under large current is included in the model in series to other two components. As a result, all fitting is excellent.

Figure 10.14 shows the measurement data (the dots) and the model fitting (curves) N+/P-well STI ESD diode I–V based on the sub-circuit given in Figure 10.12, focusing on high-current injection (TLP). The diffusion length, and finger width and finger are varied. In the model, the width of the metal interconnect and finger number is considered to achieve a scalable model.

10.3.2 Gated Diode Model

For gated diode used for ESD protection, the conventional compact diode model available in commercial simulators such as HSPICE or SPECTRE is not able to well describe the I–V characteristics because of the following two reasons: (1) the leakage current under reverse bias of gated diode caused by the gate/diffusion overlap tunneling current is larger than the junction reverse current in the conventional model; (2) there is substrate conductivity modulation associated with high injection of carriers due to mobility saturation during high-current TLP. In this section, a physics-based gated diode SPICE compact model is provided considering these two effects.

Figure 10.15 is the sub-circuit SPICE model for gated diodes in Figure 10.11. In Figure 10.15,

- Element diode is the conventional diode model.
- Element R_{tlp} is used to include the voltage-controlled resistance modulation under high-current injection (TLP).
- Element VCCS (voltage-controlled current source) is used to model the large reverse leakage current under conventional DC bias by considering the tunneling current.

The details of the element R_{tlp} and element VCCS are given as follows. Similar to the STI diode model, element R_{tlp} is used to account for conductivity modulation associated with high injection of minority carriers or mobility degradation because of high fields under TLP measurement, as detailed in Equation 10.3. Element VCCS is used to model the leakage current under reverse bias of gated diode caused by the gate/diffusion overlap tunneling current. This current (I_{VCCS}) is considered as the superposition of two currents, I_{FP} and I_{FE} (Equation 10.9), as follows. I_{FP} is related to the Frenkel–Poole emission [17],

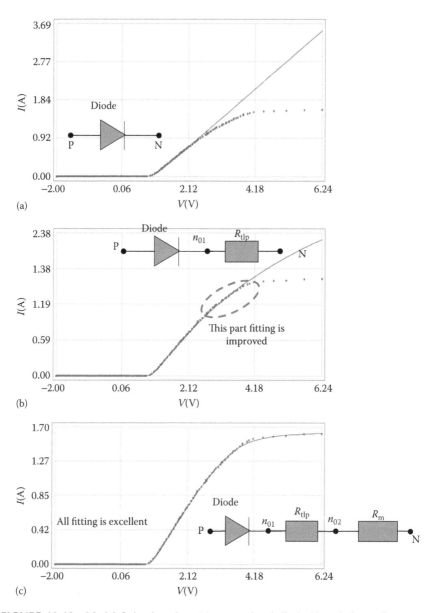

FIGURE 10.13 Model fitting based on (a) conventional diode (there is large discrepancy at high current), (b) add R_{tlp} considering substrate resistance modulation under large current, and (c) add R_m considering metal self-heating effect under large current; all fitting is excellent.

which is attributed to field-enhanced thermal excitation of trapped electrons into the conduction band and is temperature dependent (Equation 10.10). I_{FE} is related to the tunneling or field emission [17], which is caused by field ionization of trapped electrons into the conduction band or by electrons tunneling from the metal Fermi energy into the insulator conduction and is essentially independent of the temperature (Equation 10.11).

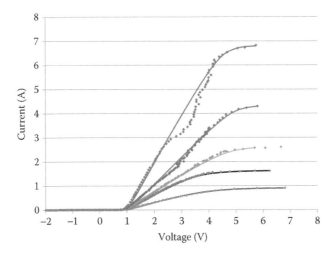

FIGURE 10.14 Fitting results of N$^+$/PW STI ESD diode I–V model at temp = 25°C.

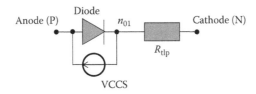

FIGURE 10.15 Sub-circuit model for the gated diode to model the I–V characteristics under both high-current TLP and low-current conventional DC bias.

$$I_{\mathrm{VCCS}} = I_{\mathrm{FP}} + I_{\mathrm{FE}} \qquad (10.9)$$

$$I_{\mathrm{FP}} = A1 \cdot W \cdot NF \cdot V_{\mathrm{d}} \cdot \exp\left(\frac{B1 \cdot \sqrt{V_{\mathrm{d}}}}{T} - \frac{q \cdot \phi_{\mathrm{b}}}{k \cdot T}\right) \qquad (10.10)$$

$$I_{\mathrm{FE}} = A2 \cdot W \cdot NF \cdot V_{\mathrm{d}}^2 \cdot \exp\left(-\frac{B2}{V}\right) \qquad (10.11)$$

where:
  $A1$, $A2$, $B1$, and $B2$ are fitting parameters
  W is finger width
  NF is number of finger
  V_{d} is the voltage across the diode
  T is temperature
  q is the electron charge
  k is the Boltzman constant
  ϕ_{b} is the barrier height for tunneling

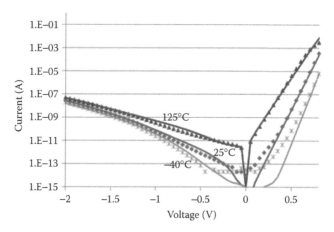

FIGURE 10.16 HP4156 tested I–V characteristics (the dots) of the P+/NW gated diode at −40°C, 25°C, and 125°C, and the corresponding modeling (the curves) based on the sub-circuit given in Figure 10.15.

It is noted that the Frenkel–Poole emission and field emission dominate at low reverse bias and high reverse bias, respectively. As shown in Figure 10.16, the leakage current under low reverse bias (0 to −1 V) is much more temperature dependent than that under higher reverse bias (−1 to −2 V). By employing Equation 10.9 to Equation 10.11 for the VCCS in the sub-circuit model shown in Figure 10.15, the model can match the silicon data very well. This is consistent with the mechanisms that I_{FP} is temperature dependent, while I_{FE} is not as given in Equations 10.10 and 10.11, respectively.

Figure 10.17 compares silicon data (the dots) and the model fitting (the curves) based on the sub-circuit given in Figure 10.15, focusing on conventional DC bias. The current is shown in log scale so that the leakage current under reverse bias can be viewed clearly. In the plot, the number of finger is varied. It is clear that the model is scalable. Thanks to the element VCCS consisting two current components (I_{FP} and I_{FE}), the excellent fitting in this regime is achieved.

Figure 10.18 compares silicon data (the dots) and the model fitting (the curves) based on the sub-circuit given in Figure 10.15, focusing on high-current injection (TLP). The current is shown in linear scale so that the large current under TLP can be viewed clearly. In the plot, the structures are the same as those shown in Figure 10.17. With the voltage-controlled resistance element R_{tlp} implemented, the model can match the silicon data very well. The model in this regime is scalable too.

In short, a physics-based new gated diode SPICE compact model is provided. The sub-circuit SPICE model includes three elements in series in-between the anode (P) and the cathode (N). The three elements are element diode for the conventional DC forward bias, element R_{tlp} to include the voltage-controlled resistance modulation under large current injection (TLP), and element VCCS to model the large reverse leakage current under conventional DC bias by considering the tunneling current. The new SPICE model matches the silicon data under both ESD TLP and normal DC forward/reverse bias very well. The model is scalable in terms of finger width and finger number.

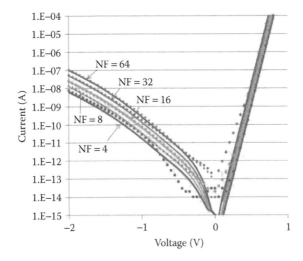

FIGURE 10.17 Silicon data (the dots) and the model fitting (the curves) based on the sub-circuit given in Figure 10.15, focusing on conventional DC bias. The current is in log scale. The gate length and the finger width are the same, while the number of finger is varied.

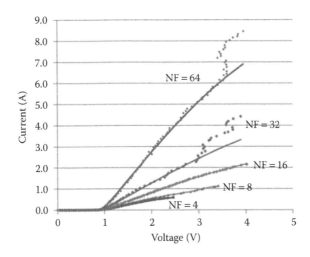

FIGURE 10.18 Silicon data (the dots) and the model fitting (the curves) based on the sub-circuit given in Figure 10.15, focusing on high-current injection (TLP). The current is in linear scale. The gate length and the finger width are the same, while the number of finger is varied.

10.4 GGNMOS MODEL

N-channel MOSFET (NMOS)-based ESD protection devices are widely used, including gate-grounded NMOS (GGNMOS) [9], gate-coupled NMOS (GCNMOS) [18], floating-body GGNMOS [19], and substrate-triggered GGNMOS [20]. It is clear that the GGNMOS is the base for other protection devices mentioned. Therefore,

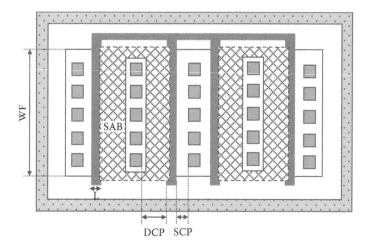

FIGURE 10.19   Schematic of the top view of the ESD GGMOS with multi-fingers.

GGNMOS model is the base for establishing the SPICE model for such protection devices. Figure 10.19 gives schematic of the top view of the ESD GGMOS with multi-fingers. The following are key layout parameters should be covered in design rule manual to ensure the desired ESD capability, including total width (W), channel length (L), drain contact to poly spacing (DCP), source contact to poly spacing (SCP), and finger width (WF). SAB (salicide block) is necessary at drain side to avoid current crowding and oxide breakdown and then increase I_{t2}. Although SCP is not important for the device reliability, it can be set to a minimum value allowed by design rule. DCP is an important design parameter for the device reliability. Large DCP can prevent current crowding and provide high reliability but consume large area. Therefore, an optimized DCP value is needed.

As the distance between the drain and source regions in deep submicron MOS devices shrinks to a value comparable to the base layer thickness in a typical BJT, the parasitic lateral BJT in the MOS device becomes prominent and can be easily activated under the ESD event. Once the parasitic BJT is turned on, the behavior of MOSFET changes dramatically. Figure 10.20 is a schematic illustration of the NMOS, its parasitic lateral BJT, and the current components during ESD stressing. In Figure 10.20a, there are five current components, among which three are electron currents (i.e., I_D, I_{II}, and I_C) and two are hole currents (i.e., I_B and I_{SUB}). I_D is the source to drain current under normal operation, which is included in the conventional SPICE model. The generation of the other current components is detailed as follows:

- In the case when $V_G = 0$ V and V_D is low (Figure 10.20b), I_D is negligible because there is a build-in barrier between the source and drain and the carriers (i.e., electrons) are hard to transport from source to drain. The corresponding I–V is provided in Figure 10.21.
- However, when V_D continuously increases to be sufficiently large, the barrier will be lowered, as shown in Figure 10.20c. Electrons are then possible

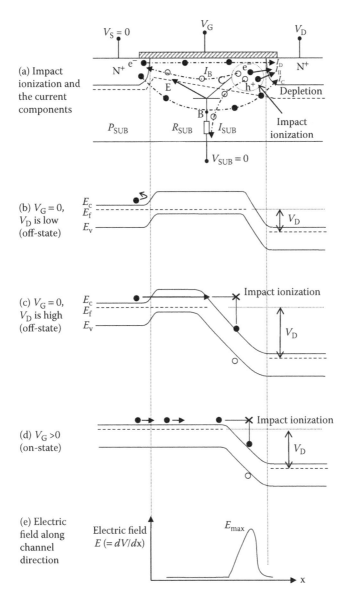

FIGURE 10.20 NMOS transistor with parasitic bipolar transistor. (a) Impact ionization and the current components; (b) VG = 0, VD is low (off-state); (c) VG = 0, VD is high (off-state); (d) VG >0 (on-state); (e) Electric field along channel direction.

tunneling from source to drain and are able to gain considerable kinetic energy when crossing the drain-substrate reverse-biased depletion region where the maximum electric field (E_{max}) is located (Figure 10.20e) and the electrons are accelerated by the electric field. These electrons, when colliding with the lattice, break covalent bonds and generate EHPs—this is

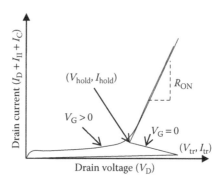

FIGURE 10.21   I–V curve with snapback behavior at $V_G = 0$ and $V_G > 0$.

so-called impact ionization. The generated EHPs may then have enough energy themselves to break more covalent bonds, thus generating even more EHPs. This phenomenon is called the *avalanche multiplication*. As shown in Figure 10.20a, the generated electrons are swept into the drain by the electric field and referred as I_{II}. I_{II} is added with I_D to form the drain current.

- On the other hand, the holes generated by impact ionization flow partly into the substrate and partly into the source because the SUB and source are both grounded and form the substrate current (I_{SUB}) and another current I_B. The electron current I_{II} mentioned earlier is the sum of the hole current I_{SUB} and I_B (i.e., $I_{II} = I_{SUB} + I_B$).
- As the semiconductor substrate has a finite resistance (R_{SUB}), the hole current gives rise to a potential drop across the source-substrate junction. In other words, the potential of node B (as shown in Figure 10.20a) will increase. As V_D increases, I_{SUB} also increases, and eventually the potential of node B becomes large enough to turn on the source–substrate junction and to cause electrons to inject from the source (i.e., the emitter of the parasitic BJT) into substrate (i.e., the base of the parasitic BJT). These electrons finally flowing to the drain area (i.e., the collector of the parasitic BJT) are referred as I_C (Figure 10.20a). The same as I_{II}, I_C is also added to the drain current. In short, the drain current includes three parts: I_D, I_{II}, and I_C.
- It is noted that the induced I_{II}, and I_C near the drain can further result in more avalanche-generated holes near the drain junction and thus a positive feedback action. When this occurs, the drain voltage is reduced quickly and maintains at a relatively constant level, a phenomenon called the *snapback*, as illustrated in Figure 10.21.
- When $V_G > 0$ V (Figure 10.20d), the channel is inverted so that I_D is much larger. There will be much more electrons involved in the impact ionization, and then the snapback will take place at a lower V_D. The corresponding I–V for $V_G > 0$ V is also provided in Figure 10.21.

In Figure 10.21, the trigger voltage V_{tr} is where avalanche multiplication first occurs and turns on the lateral parasitic bipolar transistor. Once the parasitic BJT turns on, the drain voltage is reduced and the negative resistance phenomenon is observed,

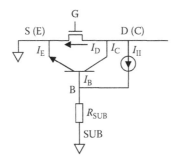

FIGURE 10.22   Equivalent circuit of the NMOS devices including parasitic lateral BJT.

because more electrons from the source drift to the drain junction until a minimum snapback voltage V_{hold} is reached. After that, the resistance is generally positive and small because of the conductivity modulation of the substrate associated with the increased current in the substrate.

The equivalent circuit of the NMOS device with the parasitic lateral BJT under ESD high-current injection is depicted in Figure 10.22 (Amerasekera's model [21]). As described earlier, the total drain current is considered to include the following three components: (1) the drain current I_D through the channel; (2) the collector current I_C of the parasitic lateral BJT; and (3) the electron current I_{II} due to impact ionization. Furthermore, I_{II} consists of the substrate current I_{SUB} passing through the substrate and the base current I_B of the parasitic BJT.

It should be pointed out that the existing NMOS model (i.e., conventional NMOS model) in SPICE does not include the components of I_{II}, R_{SUB}, and parasitic lateral BJT. For developing an NMOS model suitable for ESD simulation, these three components should be modeled and incorporated into the conventional NMOS model. R_{SUB} can be extracted from the experimental data. The expressions for I_{II}, I_B, and I_C are detailed later in this section.

As mentioned earlier, EHPs are produced near the drain junction due to impact ionization. Both the typical drain current I_D and the collector current I_C (after the parasitic BJT is turned on) can contribute to the impact ionization generated current I_{II}, as expressed as follows:

$$I_{II} = (M-1)(I_D + I_C) \tag{10.12}$$

where:
M is the avalanche multiplication factor

Prior to the turn-on of the parasitic BJT, such a current is simplified as

$$I_{II} = (M-1)I_D \tag{10.13}$$

The multiplication factor M can be modeled as [22]

$$M = \frac{1}{1 - I_{\text{ion}}} \tag{10.14}$$

where I_{ion} is the impact ionization integral:

$$I_{\text{ion}} = \int_0^{l_d} \alpha dx \tag{10.15}$$

where:
α is the impact ionization coefficient of electrons and holes
l_d is the length of the high-field region in the depletion region

α is approximated as [23]

$$\alpha = A \cdot \exp\left(\frac{-B}{E}\right) \tag{10.16}$$

where:
E is the electric field
A and B are constant parameters

The electric field typically is a function of position and has the form of $E = E_{\text{sat}}\cosh(x/ld)$ [23], where E_{sat} is the electric field with electrons under saturation drift velocity. For simplicity, E can be approximated as a constant at snapback conditions and expressed by

$$E = \frac{V_D - V_{\text{Dsat}}}{l_d} \tag{10.17}$$

Substituting Equation 10.15 through Equation 10.17 into Equation 10.14, we have

$$M = \frac{1}{1 - P_1 \exp\left[-P_2/\left(V_D - V_{\text{Dsat}}\right)\right]} \tag{10.18}$$

where:
$P_1 = Al_d$ and $P_2 = Bl_d$ are model parameters that need to be determined from measurements

The currents associated with the parasitic BJT are described as follows. The parasitic BJT is analogous to a lateral BJT, and some characteristics and features of the parasitic BJT are different from the vertical bipolar counterpart [24]. To keep the model compact, the approach of Gummel–Poon-like models [25] is not followed, but rather simple and empirical equations are used. The first-order collector current I_C can be expressed by

$$I_C = I_{\text{OC}}\left[\exp\left(\frac{V_{\text{BE}}}{V_T}\right) - \exp\left(\frac{V_{\text{BC}}}{V_T}\right)\right] \tag{10.19}$$

where:

I_{OC} is the reverse saturation current for electrons diffusing at the collector junction

$V_T = kT/q$ is the thermal voltage

k is the Boltzmann constant

T is temperature

q is the electron charge

V_{BE} is the base-emitter voltage drop

V_{BC} is the base-collector voltage drop

The base current I_B is written by

$$I_B = I_{OB}\left[\exp\left(\frac{V_{BE}}{V_T}\right) - 1\right] \tag{10.20}$$

where:

I_{OB} is the reverse saturation current for holes diffusing at the emitter junction

Both I_{OC} and I_{OB} are model parameters that need to be extracted from measurements.

Alternatively, when changing the notation to MOS (i.e., D for C and S for E), Equations 10.19 and 10.20 turn to be:

$$I_C = I_{OC}\left[\exp\left(\frac{V_{BS}}{V_T}\right) - \exp\left(\frac{V_{BD}}{V_T}\right)\right] \tag{10.21}$$

$$I_B = I_{OB}\left[\exp\left(\frac{V_{BS}}{V_T}\right) - 1\right] \tag{10.22}$$

Based on this theory, Gao et al. [26] developed a macro model of MOSFET for ESD applications consisting of the regular MOSFET, parasitic BJT, R_{SUB}, and I_{II}. The blocks for R_{SUB} and I_{II} were implemented in SPICE-like simulator Analog Hardware Definition/Description Language (AHDL).

In another effort for compact modeling of GGNMOS under ESD stressing provided by Li et al. [4], some modifications are made from Figure 10.22. This model methodology was developed using behavior language Verilog-A to describe the lateral NPN (LNPN) bipolar behavior. The equivalent circuit is given in Figure 10.23. A list of the difference is as follows:

- The avalanche multiplication factor M is separated for MOS and BJT to capture the increasing independence of drain current from gate voltage after snapback. Equation 10.12 turns to be:

$$I_{II} = \left(M_{MOS} - 1\right)I_D + \left(M_{BJT} - 1\right)I_C \tag{10.23}$$

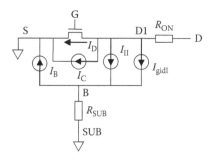

FIGURE 10.23   Equivalent circuit of the NMOS devices including parasitic bipolar mechanism.

$$M_{MOS} = 1 + \exp\left[C_1 \left(V_{DS} - C_2 - V_{Dsat} \right) \right] - \exp\left[C_1 \left(-C_2 - V_{Dsat} \right) \right] \qquad (10.24)$$

$$M_{BJT} = 1 + \exp\left[C_3 \left(V_{DB} - V_{hold} \right) \right] - \exp\left(-C_3 V_{hold} \right) \qquad (10.25)$$

where:

M_{MOS} is the multiplication factor for channel current, and it is a function of V_{Dsat}, which, in turn, is a function of V_{GS}

M_{BJT} is the multiplication factor for the parasitic bipolar current. Also note that the multiplication factor M here is different from the more typical formulation in Equation 10.18

The idea is originated from [27] to resolve the singularity problem when the denominator approaches zero in Equation 10.18.

• Drain resistance R_{ON} is included to represent the on-resistance of the GGNMOS operating in the snapback region. For simplicity, R_{ON} is set to be a constant (R_{ON}) after snapback, and it increases linearly from zero to R_{ON}. R_{ON} is a parameter to be extracted from the measurement data.

• A gate-induced drain leakage (GIDL) current (I_{gidl}) is included in the model in parallel with the impact ionization current (I_{II}). I_{gidl} is used to describe the band-to-band tunneling current commonly taking place in deep submicron technology nodes. I_{gidl} could assist to trigger the parasitic BJT if its value is large. GIDL is also considered in the model by Zhou et al. [7] as a function of V_{DS}, V_{GS}, and V_{BS} with the equation provided by BSIM4 [28].

Figure 10.24 shows the TLP data and the simulation results using the compact model given in Figure 10.23.

On the other hand, Zhou et al. [29] created a SPICE macro model for ESD MOS modeling consisting of three components, that is, an MOS transistor modeled by BSIM3 or BSIM4, a BJT modeled by VBIC (Vertical Bipolar Inter-Company), and a resistor for substrate resistance. The advantage of this macro model is that the explicit current source modeling the impact ionization (junction breakdown) in aforementioned models (Figures 10.22 and 10.23) is eliminated. This may help to increase simulation speed and reduce possibility for convergence issue.

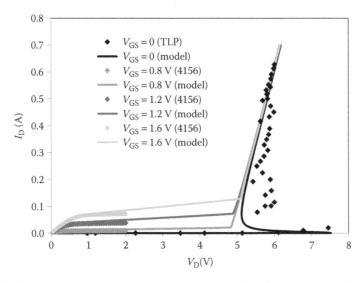

FIGURE 10.24 TLP data and the simulation results using the compact model given in Figure 10.23.

10.5 BJT MODEL

As discussed in Section 10.4, the LNPN BJT parasitic to the GGNMOS has historically served as the dominant ESD protection device. However, as semiconductor technology continues to shrink, the LNPN device has become increasingly process sensitive and less suitable for use in ESD protection [30]. Some efforts [31,32] have been made to explore vertical BJT for ESD protection.

The vertical PNP (VPNP) BJT device plays an important role in shunting ESD current to VSS (Figure 10.25). To explore a VPNP model, Torres et al. [31] extended the Standard Gummel–Poon (SGP) model using Verilog-A to include two additional

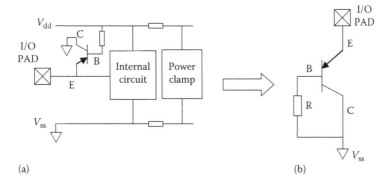

FIGURE 10.25 Electrical schematic of the vertical grounded-base PNP bipolar transistor under ESD stress (a) PNP in the ESD protection circuit and (b) equivalent circuit of the PNP for ESD protection.

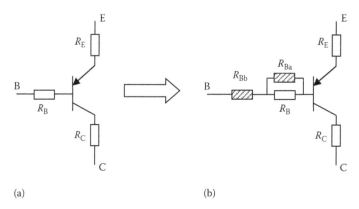

(a)   (b)

FIGURE 10.26 (a) Equivalent circuit for Standard Gummel–Poon (SGP) PNP model. (b) Extended equivalent circuit for modeling the VPNP under ESD stressing modified from SGP model.

base resistors that account for high-level injection and carrier velocity saturation, respectively, as shown in Figure 10.26.

The high-level injection effect is modeled by a voltage-dependent resistor R_{Ba}. This resistor is added in parallel to the standard SGP base resistor R_B if the base-emitter voltage (V_{BE}) exceeds the high-level injection threshold (a_0). The corresponding equation is given in Equation 10.26 [31].

$$R_{Ba} = \begin{cases} a_1 + \dfrac{a_2}{|V_{BE}| - a_0} & \left(\text{when } |V_{BE}| > a_0\right) \\ \\ \infty & \left(\text{when } |V_{BE}| < a_0\right) \end{cases} \tag{10.26}$$

The velocity saturation effect is modeled by a voltage-dependent resistor R_{Bb}. It is added in series with the standard SGP base resistor R_B if the base-emitter voltage V_{BE} exceeds the velocity saturation threshold (b_0). The corresponding equation is given in Equation 10.27 [31].

$$R_{Bb} = \begin{cases} \left(|V_{BE}| - b_0\right)^2 \cdot b_1 + \left(|V_{BE}| - b_0\right) \cdot b_2 & \left(\text{when } |V_{BE}| > b_0\right) \\ \\ 0 & \left(\text{when } |V_{BE}| < b_0\right) \end{cases} \tag{10.27}$$

On the other hand, the Ebers–Moll (EM) model was chosen by Li et al. [4] instead of the SGP model discussed earlier because it was thought that the SGP model is relatively more complex, much of which is irrelevant for ESD circuit simulation purpose. Modifications were made to account for transistor breakdown and high-current resistive effects, as shown in Figure 10.27. Compared to the equivalent circuit for EM, NPN model given in Figure 10.27a and b shows that two additional current sources [$I_{II(BC)}$ and $I_{II(CE)}$] are used to describe the impact ionization current induced by the reverse current of the collector–base junction (I_{BC}) and the emitter-injected

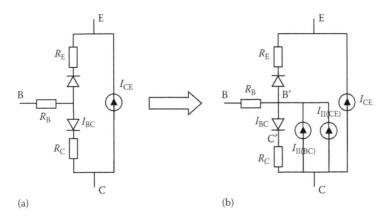

FIGURE 10.27 (a) Equivalent circuit for Ebers–Moll (EM) NPN model. (b) Extended equivalent circuit for modeling the VNPN under ESD stressing modified from EM model.

electron current (I_{CE}) when they pass through the high-field region at the collector–base junction, respectively. In both cases, the generated holes flow to the base terminal. The expressions for $I_{II(BC)}$ and $I_{II(CE)}$ are as follows:

$$I_{II(BC)} = (M-1)I_{BC} \tag{10.28}$$

$$I_{II(CE)} = (M-1)I_{CE} \tag{10.29}$$

$$M = 1 + A_1 \cdot \exp(A_2 \cdot V_{C'B'} - 1) \tag{10.30}$$

where:
A_1 and A_2 are parameters to be extracted from experimental data.

REFERENCES

1. G. Boselli, S. Ramaswamy, A. Amerasekera, T. Mouthaan, and F. Kuper, "Modeling substrate diodes under ultra high ESD injection conditions," in *Electrical Overstress/Electrostatic Discharge Symposium*, IEEE, 2001, pp. 71–81.
2. Y. Wang, P. Juliano, S. Joshi, and E. Rosenbaum, "Electrothermal model for simulation of bulk-Si and SOI diodes in ESD protection circuits," *Microelectronics Reliability*, vol. 41, no. 11, 2001, pp. 1781–1787.
3. J. Willemen et al., "Characterization and modeling of transient device behavior under CDM ESD stress," in *Electrical Overstress/Electrostatic Discharge Symposium*, IEEE, 2003, pp. 88–97.
4. J. Li, S. Joshi, R. Barnes, and E. Rosenbaum, "Compact modeling of on-chip ESD protection devices using Verilog-A," *IEEE Transactions on Computer-Aided Design of Integrated Circuits and Systems*, vol. 25, no. 6, 2006, pp. 1047–1063.
5. Z.H. Gan, W.S. Wong, and J.J. Liou, "Electrostatic discharge protection of integrated circuits," in *Semiconductor Process Reliability in Practice*, McGraw-Hill, New York, 2013, Chapter 9.

6. A. Amerasekera, C. Duvvury, W. Anderson, H. Gieser, and S. Ramaswamy, *ESD in Silicon Integrated Circuits*, 2nd Edition, John Wiley & Sons, England, 2002.

7. Y.Z. Zhou, J.-J. Hajjar, and K. Lisiak, "Compact modeling of on-chip ESD protection using standard MOS and BJT models," in *Proceedings of the 8th International Conference on Solid-State and Integrated Circuit Technology*, IEEE, 2006, pp. 1–4.

8. M. Mergens, W. Wilkening, S. Mettler, H. Wolf, and W. Fichtner, "Modular approach of a high current MOS compact model for circuit-level ESD simulation including transient gate coupling behavior," in *Proceedings of the 37th International Reliability Physics Symposium*, IEEE, 1999, pp. 167–178.

9. C. Russ, K. Bock, P. Roussel, G. Groseneken, H. Maes, and K. Verhaege, "A compact model for the grounded-gate nMOS behavior under CDM ESD stress," in *Electrical Overstress/Electrostatic Discharge Symposium*, IEEE, 1996, pp. 302–315.

10. H. Anzai, Y. Tosaka, K. Suzuki, T. Nomura, and S. Satoh, "Equivalent circuit model of ESD protection devices," *Fujitsu Scientific Technical Journal*, vol. 39, 2003, pp. 119–127.

11. V. Puvvada, V. Srinivasan, and V. Gupta, "A scalable analytical model for the ESD N-well resistor," in *Electrical Overstress/Electrostatic Discharge Symposium*, IEEE, 2000, pp. 437–445.

12. A. Amerasekera, M.-C. Chang, J. Seitchik, A. Chatterjee, K. Mayaram, and J.-H. Chern, "Self-heating effects in basic semiconductor structures," *IEEE Transactions on Electron Devices*, vol. 40, no. 10, 1993, pp. 1836–1844.

13. K. Banerjee et al., "High-current failure model for VLSI interconnects under short-pulse stress conditions," *IEEE Electron Device Letters*, vol. 18, no. 9, 1997, pp. 405–407.

14. T. Maloney, "Integrated circuit metal in the charge device model: bootstrap heating, melt damage, and scaling laws," in *Electrical Overstress/Electrostatic Discharge Symposium*, IEEE, 1992, pp. 129–134.

15. J.R. Manouvrier, P. Fonteneau, C.A. Legrand, C. Richier, and H. Beckrich-Ros, "A scalable compact model of interconnects self-heating in CMOS technology," in *Electrical Overstress/Electrostatic Discharge Symposium*, IEEE, 2008, pp. 88–93.

16. M.-T. Yang, Y. Du, C. Teng, T. Chang, E. Worley, K. Liao, Y.-W. Yau, and G. Yeap, "BSIM4-based lateral diode model for RF ESD applications," in *IEEE 11th Annual Wireless and Microwave Technology Conference*, IEEE, 2010, pp. 1–5.

17. S.M. Sze, *Physics of Semiconductor Devices*, 2nd Edition, John Wiley & Sons, New Jersy, 1981.

18. M.X. Huo, K.B. Ding, Y. Han, S.R. Dong, X.Y. Du, D.H. Huang, and B. Song, "Effects of process variation on turn-on voltages of a multi-finger gate-coupled NMOS ESD protection device," in *Proceedings of the 16th IEEE International Symposium on the Physical and Failure Analysis of Integrated Circuits*, IEEE, 2009, pp. 832–836.

19. J.-I. Won, J.-W. Jung, I.-S. Yang, and Y.-S. Koo, "Design of ESD protection device using body floating technique in 65 nm CMOS process," *Electronics Letters*, vol. 47, no. 19, 2011, pp. 1072–1073.

20. M.-D. Ker and T.-Y. Chen, "Substrate-triggered ESD protection circuit without extra process modification," *IEEE Journal of Solid-State Circuits*, vol. 38, no. 2, 2003, pp. 295–302.

21. A. Amerasekera, M.-C. Chang, C. Duvvury, and S. Ramaswamy, "Modeling MOS snapback and parasitic bipolar action for circuit-level ESD and high-current simulations," *IEEE Circuits and Devices Magazine*, vol. 13, no. 2, 1997, pp. 7–10.

22. S.E. Laux and F.H. Gaensslen, "A study of channel avalanche breakdown in scaled n-MOSFET's," *IEEE Transactions on Electron Devices*, vol. ED-34, no. 5, 1987, pp. 1066–1073.

23. Y.A. El-Mansy and A.R. Boothroyd, "A simple two-dimensional model for IGFET operation in the saturation region," *IEEE Transactions on Electron Devices*, vol. ED-24, no. 3, 1977, pp. 254–262.

24. S. Verdonckt-Vandebroek and S.S. Wong, "High-gain lateral bipolar action in a MOSFET structure," *IEEE Transactions on Electron Devices*, vol. 38, no. 11, 1991, pp. 2487–2495.

25. G. Massobrio and P. Antognetti, *Semiconductor Device Modeling with SPICE*, 2nd Edition, McGraw-Hill, New York, 1993.

26. X.F. Gao, J.J. Liou, J. Bernier, G. Croft, and A. Ortiz-Conde, "Implementation of a comprehensive and robust MOSFET model in Cadence SPICE for ESD applications," *IEEE Transactions on Computer-Aided Design of Integrated Circuits and Systems*, vol. 21, no. 12, 2002, pp. 1497–1502.

27. S.L. Lim, X.Y. Zhang, Z. Yu, S. Beebe, and R.W. Dutton, "A computationally stable quasi-empirical compact model for the simulation of MOS breakdown in ESD-protection circuit design," in *International Conference on Simulation of Semiconductor Processes and Devices,* IEEE, 1997, pp. 161–164.

28. W. Liu, *MOSFET Models for SPICE Simulation Including BSIM3v3 and BSIM4*, John Wiley & Sons, New York, 2001.

29. Y. Zhou, D. Connerney, R. Carroll, and T. Luk, "Modeling MOS snapback for circuit-level ESD simulation using BSIM3 and VBIC model," in *Sixth International Symposium on Quality of Electronic Design*, IEEE, 2005, pp. 476–481.

30. E. Worley, A. Salem, and Y. Sittampalam, "High current characteristics of devices in a 0.18 μm CMOS technology," in *Electrical Overstress/Electrostatic Discharge Symposium*, IEEE, 2000, pp. 296–307.

31. C.A. Torres, J.W. Miller, M. Stockinger, M.D. Akers, M.G. Khazhinsky, and J.C. Weldon, "Modular, portable, and easily simulated ESD protection networks for advanced CMOS technologies," in *Electrical Overstress/Electrostatic Discharge Symposium*, IEEE, 2001, pp. 81–94.

32. G. Bertrand, C. Delage, M. Bafleur, N. Nolhier, J.-M. Dorkel, Q. Nguyen, N. Mauran, D. Trémouilles, and P. Perdu, "Analysis and compact modeling of a vertical grounded-base n-p-n bipolar transistor used as ESD protection in a smart power technology," *IEEE Journal of Solid-State Circuits*, vol. 36, no. 9, 2001, pp. 1373–1381.

11 Advanced TCAD Methods for System-Level ESD Design

Vladislav A. Vashchenko and Andrei A. Shibkov

CONTENTS

11.1 TRADITIONAL AND PARAMETERIZED TCAD APPROACHES

11.1.1 PROCESS TECHNOLOGY DEVELOPMENT AND ESD IP DESIGN WITH TCAD

Technology computer-aided design (TCAD) is a branch of electronic design automation that models semiconductor devices fabrication and operation. It relies on physical process and device numerical simulation methods applied to integrated process technologies for integrated circuit (IC) manufacturing. Numerical solution of finite-element models (FEMs) combines several steps accomplished with specialized physical simulation tools, output data visualization, and simulation flow management. In analog design with 0.5 μm–40 nm technology nodes, the majority of the simulation problems can be adequately addressed within two-dimensional (2D) approach. Three-dimensional (3D) simulations are randomly used, unless essentially

3D devices, for example, FinFETs are analyzed. Process simulation method is applied to obtain the device regions with diffused doping profiles using numerical solution of the physical models for the process steps of deposition, etching, implantation, annealing, oxidation, and epitaxial growth. The input parameters for the process simulation steps mimic the fabrication tool settings. Electrical parameters of FEM devices are obtained with device simulation tools after FEM device model is constructed using the corresponding donor and acceptor density profiles, electrodes, and proper simulation mesh.

TCAD engineers in process development organizations are mainly focused on the physical process simulation to create the calibrated process simulation flow. This flow is expected to adequately predict the electrical parameters of multiple integrated devices represented by rather standard architectures. The device simulation during new technology integration remains rather simple. In contrast, in case of electrostatic discharge (ESD) Intellectual Property (IP) development for analog processes and especially system-level design, the focus is shifted on the device circuit-level solutions within constraints of the already "frozen" semiconductor technology. The solutions based on device circuit co-optimization represent a preferred approach to address the development needs without additional process steps or mask layers. ESD protection development must take into account both the transient nature of the ESD pulses and complex interactions between the clamp and internal circuit in nonlinear conductivity modulation modes with high rate of impact ionization and injection. This necessitates that transient coupled circuit device (or mixed-mode) analysis is used as the main simulation approach in the ESD design methodology.

ESD engineers can rarely afford to focus full time on traditional TCAD analysis in multi-tasking environment with diverse responsibilities and aggressive timelines for delivering the working solutions. While new process technology platform development cycle usually lasts for almost three years, ESD IP design or problem fixing often has to be accomplished within weeks. Under these conditions, the accuracy of numerical analysis can often be sacrificed to some extent in favor of speed. This not only requires a novel TCAD methodology for the ESD design but ultimately dictates a set of requirements for a new type of software tool with the architecture and simulation flow different from traditional tools. The overview of the specific tool requirements and the best practices is presented here followed by the demonstration of the new TCAD methodology on the examples typical for system-level analog ESD design.

11.1.2 Parameterization and Automation in TCAD

The features of traditional TCAD tools relevant for analog ESD design problems have hardly evolved over the past two decades. The tool usability user friendliness-oriented features remain rather limited because of the tool architecture carryovers from the original research-oriented university code created at the dawn of the computer age. A set of independent software tools still requires extensive user intervention during the intertool communication and data export–import for visualization and analysis. In spite of the recent advances in GUI design, the TCAD tools remain mainly script-driven and very difficult in practical use for an engineer without extensive TCAD background. An engineer without decades of experience in specific vendor TCAD tools can hardly expect to produce any reliable simulation results in a reasonable amount of time. Moreover, the key mixed-mode capability required for ESD design was introduced in traditional TCAD tools as an add-on and hardly any effort went into making the mixed-mode simulation features practically usable. Thus, for regular ESD and IC design engineers, the traditional TCAD tool set gives an impression equivalent to the comic version of Swiss army knife (Figure 11.1) where multiple potentially useful features are available but extremely hard to use. The tools can simulate some aspects of exotic technologies (nanotechnologies, GaN, etc.) but are hard to apply for standard Si semiconductor device design, even when the calibrated simulated process flow is in place.

Meantime, the majority of the ESD and especially system-level problems remain within the scope of Si power analog processes in extended 0.13–0.5 µm CMOS and BCD technology nodes with integrated lateral high-voltage devices. The challenges are related to the system-level ESD protection IP design, latch-up, ESD network, and internal circuit interaction. Addressing these challenges with an empirical design approach is hardly efficient. Before the age of widely available personal computers and workstations, the semiconductor engineers were mainly applying the analytic approach to gain physical understanding for semiconductor device design. Such analytical approach was productive for the semiconductor components with rather high feature dimensions. After decades of aggressive scaling in modern technologies, an adequate analytical description of the integrated devices in ESD operation conditions with nonlinear electrothermal effects is hardly possible. When new to the field, engineers are relying on empirical approach mimicking those who have many years of experience in the field, the created methodological gap, leads to

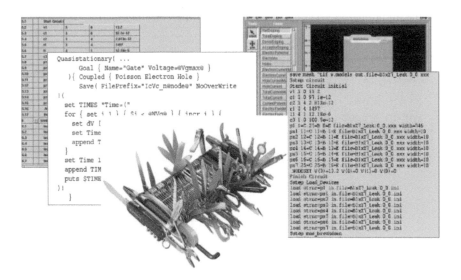

FIGURE 11.1   The user interface of the traditional TCAD tools represents a real puzzle for engineers without extensive TCAD background.

design mistakes, increases the number of design cycles, and can result in missing the design targets. Thus, application of TCAD for ESD design is not a luxury but a critical necessity [1].

Unlike conventional circuit simulation with compact models, the mixed-mode simulation combines both an accurate solution of the semiconductor model equations in the FEM devices in nonlinear conductivity modulation regimes and circuit elements represented by compact models. Mixed-mode simulation already became the key approach for the ESD design. It consists of solving the circuit equations coupled with simultaneous solution of the carrier transport in FEM devices in transient time domain (Figure 11.2). The complexity of the circuit that can be analyzed in mixed mode is determined by total number of mesh points in FEM devices. However, increased productivity of computer systems allows mixed-mode analysis of many practically important cases specific to ESD design including complex system-level cases.

The mixed-mode simulation approach can also be implemented in traditional TCAD approach. However, the architecture of the traditional TCAD tools significantly limits the complexity of problems that can be practically addressed. For mixed-mode simulation with traditional TCAD approach, the simulation flow is defined in terms of single-point tools (Figure 11.3a). Every FEM device in the final mixed-mode circuit first requires full-process simulation for a fixed set of device parameters. Then DC electrical characteristics have to be obtained using a separate device simulator. A script for the mixed-mode simulation circuit must be written including the references to the files for each FEM device. Finally, after transient mixed-mode simulation is performed, the solution has to be analyzed using separate visualization tools (Figure 11.3a).

Any new change in the device structure parameters requires re-running the process simulation first, followed by re-meshing of the FEM structures, and then re-running the mixed-mode circuit simulation. Clearly, the mixed-mode analysis with the traditional TCAD flow is an extremely time-consuming effort. It involves a

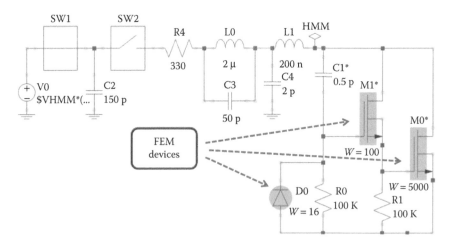

FIGURE 11.2　Mixed-mode simulation circuit for high-voltage active clamp with HMM ESD pulse.

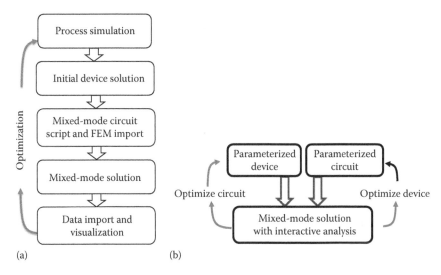

FIGURE 11.3   Mixed-mode simulation flow with traditional TCAD (a) and parameterized (b) tools.

substantial work effort to overcome convergence issues specific to complex mesh of large devices. The work bench can control multiple simulations with defined input in already-refined simulation flow for the process parameter splits, but it hardly can enable any mixed-mode analysis automation.

A completely different mixed-mode simulation flow (Figure 11.3b) is realized with a specialized tool DECIMM™ (*DE*vice-*CI*rcuit *M*ixed-*M*ode) [2]. It combines all the necessary simulation tools with interactive GUI and built-in data management, post-processing, and visualization tools in a self-contained package. DECIMM™ is not simply a workbench for setting up and managing the simulation runs but a powerful self-contained tool designed as an efficient and flexible mixed-mode simulator. Although the FEM structures created by process simulators can be imported into DECIMM™ too, this approach limits the benefits of the complete parameterization of the device regions and doping profiles.

Creation of the parameterized device templates and parameterized process descriptions for each technology is a one-time investment (Figure 11.3b), which enables a new level of design automation. Using the parameterized approach, the FEM structure with desired parameters is automatically "assembled" at the beginning of each mixed-mode simulation run using the defined desired parameter values. The current transport equations in the FEM devices are solved together with the circuit equations during the transient mixed-mode simulations, thus providing the highest degree of numerical stability and practically guaranteeing convergence in all physically valid cases.

The combination of interactive GUI for the input and analysis of the simulation results, device templates with parameterized regions and doping profiles, extensive automation of simulation setup, and data processing offers an entirely new level of simulation efficiency [3]. A mixed-mode simulation with such self-contained mixed-mode simulator can be performed once the parameterized doping profiles and device templates are defined. After one-time investment into building, the parameterized device templates and process description of any simulation can be immediately performed by simply changing any of the circuit, compact model, device, process, or physical model parameters. This qualitatively new level of flexibility allows specifying an algorithmic link between the input parameter values and numerical simulation results to define and carry out additional simulations. This new approach has already been applied to TLP, latch-up, process capability index, and IC system co-design [3,4].

11.1.3 PARAMETERIZED DEVICE TEMPLATE

The parameterized device template contains all the information about the device regions geometrical parameters, mesh, and analytical doping profiles. For example, the NMOS device template (Figure 11.4) is defined using a simple sequence of the device region length parameters in horizontal (X) direction from left to right for the lengths of drain LD, spacer Lsp, poly gate LP, source LS, body-to-source shallow trench isolation $Lsti$, and body LB. In vertical (Y) direction, the corresponding region thickness parameters $Tpoly$, Tox, $Tsti$, and Tsi are used to define thickness of the poly, gate oxide, shallow trench isolation, and bulk Si, respectively (Figure 11.4b). The NMOS region boundary description defined in the template requires only 18 points with assigned parametric coordinates, as shown in the interactive device boundary editor window (Figure 11.4a). The template defines all regions, electrodes, doping profiles, and mesh in the device (Figure 11.4b). For example, to define drain implant and electrode, the derived variables $DrainL$ and $DrainR$ are introduced to define the left and right edges of the drain region. $DrainR$ is simply calculated by adding the drain length parameter value LD to the coordinate $DrainL$. The total structure length is likewise obtained by a summation of all regions lengths: $Lmax = LD + Lsp + LP + Lsp + LS + Lsti + LB$. Thus, a completely parameterized template is created in which changes in the value of one parameter automatically adjust the values of all the related parameters keeping the structure regions, mesh, and doping profiles self-consistent.

In spite of some necessary simplification of the regions shape and doping profiles, the accuracy of the numerical analysis remains adequate. The simplifications introduced in the process of creating the parameterized templates are compensated by the flexibility of the approach that allows for interactive analysis with real-time physical parameters variation. While the exact doping profiles still can be imported from process simulation or characterization data, the analytical approach using doping distributions described in terms of error functions in lateral and Gaussian functions in vertical directions was found to be adequate.

The new parameterized tool architecture enables several ways of the FEM device template synthesis (Figure 11.5). In addition to interactive template definition, the template can be automatically extracted from the layout simply by drawing a cutline as outlined in Section 11.1.5. The most convenient way of handling the doping profile information is creation of the parameterized process file for the entire process.

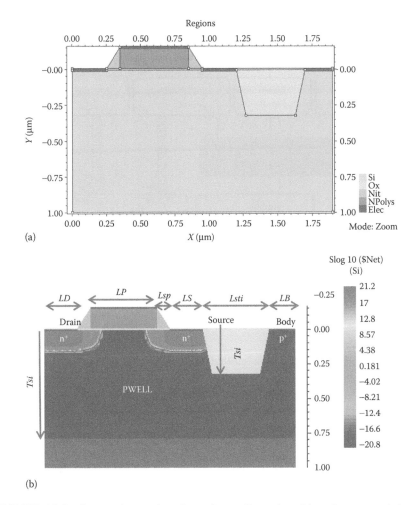

(a)

(b)

FIGURE 11.4   Interactive regions boundary editor view (a) and generated device cross section (b) for NMOS.

11.1.4 DIFFERENT OPTIONS OF CREATING THE PARAMETERIZED FEM DEVICE TEMPLATE

Extraction of the analytical doping profile parameters can be accomplished in several ways [2]. The most straightforward approach is using the physical process simulation to obtain the diffused profiles for the entire process. Although the profiles can be extracted directly from the process simulation results for the entire FEM device structures, the most accurate approach is based on the single-mask process simulation. Unlike in traditional TCAD approach, in this case the process simulation is used only once just to obtain doping profiles under the shallow trench isolation and open Si surface. Then an interactive curve fitting tool is used for the extraction of the

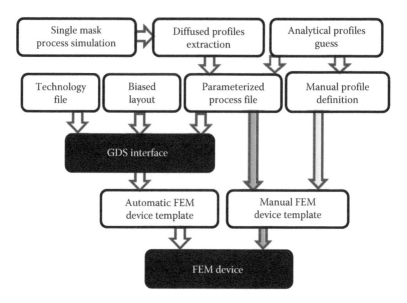

FIGURE 11.5   Options to create the parameterized FEM device.

analytic implant profiles from the cut lines in *x*- and *y*-directions. For example, the interactive fitting procedure applied to the N-well profile (Figure 11.6a) produces a Gaussian profile with two peaks and corresponding characteristic *Ych* parameters. Similarly, the characteristic error function *Xch* parameters are extracted using the horizontal cutlines. This procedure is used one time for every relevant implant in the process to create parameterized description for the entire process (Figure 11.6b), and the doping profile parameters are complemented by the region parameters (Figure 11.6c). The region parameters are used when creating the FEM device templates directly from the layout as described in Section 11.1.5.

When the calibrated TCAD process simulation flow is unavailable, the doping profile information can be requested directly from foundry or a test chip with structures for SIMS profile extraction can be fabricated and characterized. In many cases, an acceptable accuracy can be obtained by using a best guess for the doping profiles based on the information available for the similar technologies and adjusting the profiles by comparing the simulation results with the available electrical test data for the devices fabricated using the process in question. This way of process representation enables generating the FEM models for every possible device in the integrated process. The process parameters are varied in the mixed-mode simulation projects similarly to the device geometry and circuit model parameters. Therefore, a flexible algorithm can be applied to generate the input parameters for the new run based on previous run results.

It has been experimentally verified that parameterized FEM devices models with analytical doping profiles offers simulation accuracy comparable to that using the device structures created by the process simulation. The superposition of the analytical representation of the diffused doping profile components allows adequate

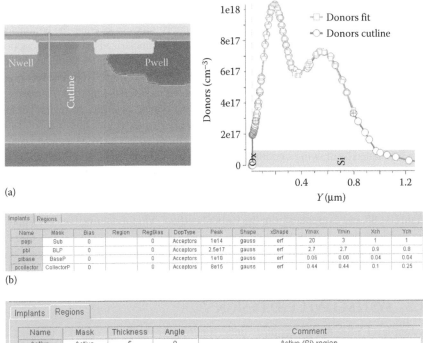

(a)

(b)

Name	Mask	Bias	Region	RegBias	DopType	Peak	Shape	xShape	Ymax	Ymin	Xch	Ych
pepi	Sub	0		0	Acceptors	1e14	gauss	erf	20	3	1	1
pbl	BLP	0		0	Acceptors	2.5e17	gauss	erf	2.7	2.7	0.9	0.8
pibase	BaseP	0		0	Acceptors	1e18	gauss	erf	0.06	0.06	0.04	0.04
pcollector	CollectorP	0		0	Acceptors	8e15	gauss	erf	0.44	0.44	0.1	0.25

(c)

Implants	Regions			
Name	Mask	Thickness	Angle	Comment
Active	Active	5	0	Active (Si) region
STI	-	0.3	7	STI (Ox) region
Poly	Poly	0.2	0	Poly (NPolyS or PPolyS) region
Gox	-	0.0131	0	Gate Oxide (Ox) region
Contact	Contact	0	0	Contact region
Spacer	-	0.02	0	Gate sidewall spacer region
GoxII	Gox	0.0031	0	LV Gate Oxide (Ox) region defined by Gox mask

FIGURE 11.6 A two-peak diffused profile from a single-mask process simulation with extracted analytical profile (a), a section of parameterized process description with the extracted profile parameters (b), and device regions description from the same parameterized process file (c).

recreation of all the devices in the integrated process. Approximating the vertical doping profiles with Gaussian and lateral with error functions provides an adequate analytical approximation of the actual doping profiles relevant to the analog processes.

The major benefits of the new approach include efficient analysis of very large FEM device structures required to study the latch-up phenomena and multi-finger array cells; automated statistical analysis; high-current system-level ESD in power-on conditions; surge analysis with electrothermal effects, and interactive real-time optimization. The application of the new approach using parameterized devices is explained in the following examples after description of the basic principles of the system-level ESD design.

11.1.5 Layout Interface for Automated FEM Device Template Synthesis

The layout interface creates a spectrum of new opportunities to leverage the parameterized template-based simulation approach [2]. The parameterized process description already contains all necessary information to generate both the FEM device cross-sections and the parameterized templates directly from the layout. To enable this functionality, additional technology-specific information is required to link the layer numbers with mask names and types and the corresponding doping profiles and region parameters. The mask types are used to identify the implants and regions (the silicide block, poly, spacer, dual gate oxides, and STI isolation) (Figure 11.7). The FEM template generation is initiated by drawing a cutline across any region of interest in the imported layout.

The example FEM cross sections generated automatically directly from the layout of a grounded-gate snapback NMOS cell show the double gate NMOS device created from the horizontal cutline (Figure 11.8a and c) and the full lateral isolation junction (Figure 11.8b and d) created from the vertical cutline. After assigning the corresponding electrode names, both FEM devices can be simulated to obtain various electrical characteristics such as the drain-source I–V characteristics for different gate bias (Figure 11.8e) and the breakdown characteristics of the isolation region (Figure 11.8f).

11.2 SYSTEM-LEVEL ESD DESIGN

System-level ESD design requires understanding of the pulse regimes realized at standard ESD and surge pulses, protection principles, and on-chip clamps. It is greatly assisted by the advanced parameterized mixed-mode TCAD analysis. In this section, system-level ESD subject is summarized up to a level required for understanding of the related mixed-mode analysis. A more detailed and structured presentation can be found in [5]. Systems can experience real-life-level ESD, electromagnetic interference (EMI), and surge overvoltage events of much higher level than possible

Layer	Name	Visual	Type	Implant	Comment
2:0		✔	Active		Active
4:0	POL	✔	Poly	N+	poly Si
5:0	NP	✔	Implant	N+	Nplus
6:0	PP	✔	Implant	P+	Pplus
7:0	CNT	✔	Contact		
8:0	M1	✔			Metal1
9:0	Via	✔			Via12
10:0	M2	✔			Metal2
25:0	SB	✔	SiBlock		Silicide Blk
45:0	PW	✔	Implant	PwellHV	Pwell
Wafer			Implant		Blanket imp

Technology tabs: Hierarchy | Technology | Cutline
Technology: BCD018
Process: BCD018

FIGURE 11.7 Technology panel view in the layout interface.

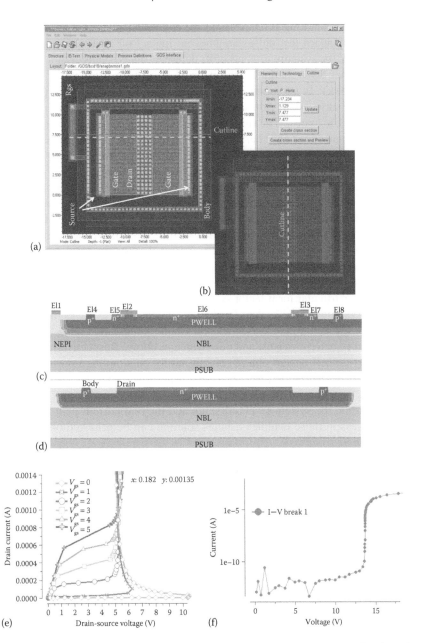

FIGURE 11.8 Layout interface view for parameterized device template extraction from the layout view with horizontal cutline (a) corresponding to the dual gate NMOS device (c) and vertical cutline (b) corresponding to the body diode (d). The simulated I_d-V_d (e) and body diode breakdown (f) characteristics.

component-level handling ESD events in the controlled environment. Specification to withstand standard test pulses assumes no impact on long-term reliability and functionality of the systems as a part of IC datasheet and systems certification. Real-life environment ESD charge can be induced by voltage levels of the conductive objects up to 20 kV. In addition to ESD events, different fast EMI, slow overvoltage, and secondary surge events can generate a significant overstress in the time domains different from ESD pulse. The surge pulses can be generated due to load mismatch, intermittent connections and short circuits, load dump (e.g., sudden power car stereo system switching off), lightning strike, or defibrillator pulses as well as a misuse of the systems, for example, above the absolute maximum ratings.

Key differentiating aspects of the system-level on-chip ESD design include pulse waveforms, much higher-current amplitude, and sustainability at the power-on conditions. The system-level standards were originated specifically for the systems rather than for IC pins interfacing with system-level ports. A methodological gap between the ESD design of the systems and on-chip system-level pins generates mistreating of the qualification test standards applied to evaluate IC pin capability, miscorrelation between the ESD passing level of stand-alone protection cell on the test chip, and particular form-factor systems with additional PCB components.

Only very roughly the system-level ESD specification at the component level can be assumed as a higher magnitude current under different pulse waveform to be sustained in both power-off and power-on conditions. Respectively, the on-chip ESD protection cell design objectives can hardly be simplified down to a width scaling of the conventional protection cells. The waveforms and the current level at both the system-level IC pins and adjacent analog domains are significantly dependent upon the system-level test setup and the real current path in particular IC. A progress in these complex aspects solution can be achieved only using the mixed-mode analysis.

11.2.1 System-Level Test Methods

The most broadly used system-level ESD pulse standards are the IEC 61000-4-2 [6] for general electrical equipment and ISO 10605 [7] for electronic modules for vehicle (Table 11.1). Both standards define similar double-peak waveforms (Figure 11.9a) with the current parameters as linear function of the pre-charge voltage, number of zaps and details of test setup for the contact, and air-gap discharges applied from ESD gun (Figure 11.9b).

Perhaps only the contact discharge waveform is more or less defined within the wide variation of the pulse parameters and can be represented by an equivalent setup with the so-called human metal model (HMM) pulse [8]. HMM mimics a bench setup with ESD gun-like pulse generator connected directly to IC pins either in the package or on the wafer. The air-gap waveforms are more complex and setup dependent (Figure 11.10).

The load-dependent standard surge pulse can be produced by the combination wave generator with optional coupling–decoupling network. The surge standard [9] is used for on-chip design specification bringing the pulse duration to much higher 8–50 µs level [10,11]. In this time domain, the self-heating of the ESD cells cannot

TABLE 11.1

Comparison of the Main Features of IEC 61000-4-2 and ISO 10605 Standards

Standard	IEC 61000-4-2	ISO 10605
Target	General electrical equipment	Electronic modules for vehicle
Preferred test	Contact discharge method	Air discharge method, direct ESD, powered DUT
RC network	150 pF; 330 Ω	150/330 pF, 330 Ω; 150/330 pF, 2 kΩ
Gun ground connection	To GRP through $R = 970$ kΩ	Direct: powered DUT HCP and DUT GND Indirect: powered DUT HCP or GRP
Preferred levels (kV)	Contact: 2, 4, 6, 8 Air: 2, 4, 8, 15	Direct: Contact: 2–8; 4–15; air: 2–15; 4–15; 6–25 Indirect: 2–8; 2–15; 4–20
Number of zaps	10 single in the most sensitive polarity	Direct, unpowered, or vehicle test method: 3 discharges for each test voltage and polarity Indirect: 50 for each test voltage and polarity

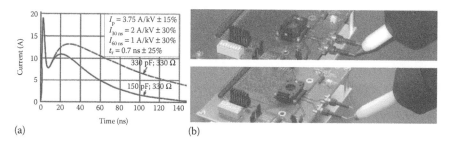

(a)   (b)

FIGURE 11.9 Comparison of the IEC 61000-4-2 and ISO 10605 standard waveforms (a) and examples of the contact and air-gap discharge gun testing on PCB (b).

be neglected. In simulation it has to be analyzed for significantly large FEM device cross sections in comparison with small adiabatic region in case of component-level ESD pulses. A significant miscorrelation between the current level of the component and surge stress (Table 11.2) is the result of the electrothermal phenomena due to high surge pulse energy (Table 11.3). Therefore, the physical effects in both the front-end Si structure and the back-end metallization have to be taken into account at mixed-mode simulation analysis with the corresponding equivalent circuits for pulse sources (Figure 11.11a through c). Certainly, understanding and comparison with the experimental results has to be applied to account for the physical limitations due to 3D effects of current distribution, calibration of model coefficients, and parasitic current path formed in actual chip layout. Comparison of the simulated pulse waveforms for these circuits (Figure 11.11d) demonstrates substantially higher energy dissipated at system-level stress (Table 11.3) in comparison with the component pulses.

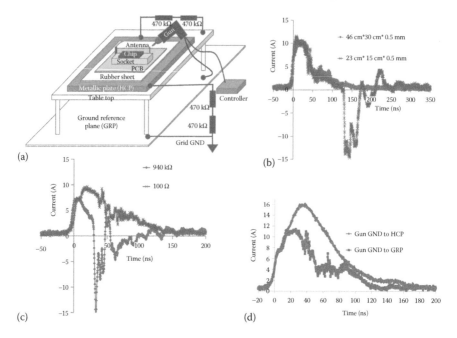

FIGURE 11.10 Setup (a) and 8 kV air-gap discharge waveforms for different PCB plate size (b), bleeding resistors (c), and gun ground connection (d).

TABLE 11.2
ESD and Surge Passing-Level Miscorrelation for Typical ESD Structures

POSITIVE (Snapback)	TLP IT2 (A)	HBM (A)	HMM (A)	Negative Surge (A)	Positive Surge (A)
5 V snapback NMOS	4.6	4.6	6	0.45	
60 V DIAC	>15			4	3.5
40 V NLDMOS-SCR	>15			>4	>4

TABLE 11.3
Comparison of the Energy of Component and System-Level ESD and Surge Pulses

Pulse	C (pF)	V_{pulse}	$E = CU^2/2$
IEC surge	6038000	100 V	30.2 mJ
IEC cont	150	8 kV	4.8 mJ
IEC air	120	15 kV	13.5 mJ
MM	220	200	4.4 µJ
HBM	100	2 kV	200 µJ
CDM	5	0.75 kV	1.4 µJ

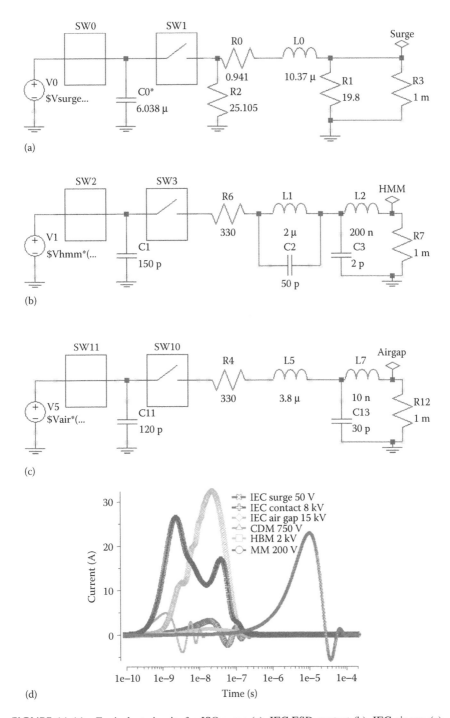

(a)

(b)

(c)

(d)

FIGURE 11.11 Equivalent circuits for ISO surge (a), IEC ESD contact (b), IEC air gap (c), pulses and simulated waveforms for the short circuit load (d).

11.2.2 SYSTEM-LEVEL ESD NETWORK PROTECTION PRINCIPLES

The main approach of the system protection is to separate the system ports and active IC components by two-stage network. When the port is not directly attached to IC pin in high-current conditions, the electric potential of the port must be limited to avoid isolation damage or EMI of the system blocks. This is achieved by local port protection with transient voltage suppressors (TVS). As on-chip protection cannot sufficiently limit down the voltage at the remote port (Figure 11.12), a typical design includes on-PCB TVS practically at every port and antenna (Figure 11.13). The protection principles of the systems are similar to IC components: (1) Local port protection with TVS is similar to local clamps (Figure 11.14a), (2) the rail-based multi-port TVS module is similar to core clamp and diodes in IC (Figure 11.14d), (3) self-protection of the components in case of really small form-factor systems when the port is physically attached to the IC pin, and (4) multi-stage network for PCB and IC components.

Implementation of a two-stage protection network is realized with the first-stage discrete TVS (Figure 11.14a through c). The second-stage discharge current path

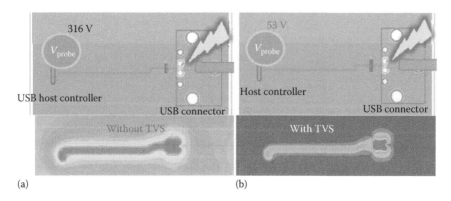

(a)   (b)

FIGURE 11.12 Port protection effect with (a) and without (b) TVS.

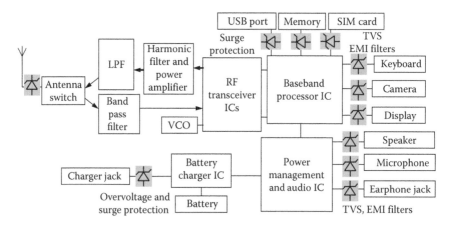

FIGURE 11.13 Cell phone handset protection network.

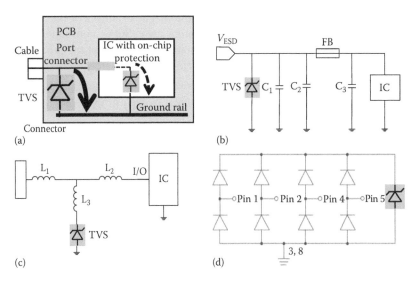

FIGURE 11.14 Two-stage protection principle for systems (a), ESD networks with transient voltage suppressor (b, c), and multi-port TVS module (d).

is provided by additional PCB components including on-chip clamps. The stage separation network is accomplished with distributed capacitance and inductance (Figure 11.14b and c).

High-speed signal and portable power sources do not allow substantial filtering of the signal waveforms or power losses. This limits impedance path between the port and the IC pin, thus increasing the high-current and precision requirements for the system-level ESD solutions. It generates trend toward much higher-current capability, low-parasitic on-chip protection components and require more precise Si TVS components.

In combination with portable devices form-factor reduction, a single stage on-chip protection of the system port eliminates TVS application cost too when IC pin is connected close to the system port. Alternatively, an IC system co-design approach is used to match the pin protection capability with remaining system primary protection network. Many aspects of such co-design can be addressed with mixed-mode TCAD analysis to obtain coupled solution of the on-PCB and on-chip ESD networks.

11.3 MIXED-MODE SIMULATION CASES FOR SYSTEM-LEVEL ESD DESIGN

Automated parameterized analysis can be used to solve a broad spectrum of problems that are difficult to address with traditional TCAD tools due to unfavorable cost–benefit ratio, convergence issues, and excessive simulation time. In the parameterized process-device approach, the results from the previous mixed-mode simulations can be used for algorithmic control and automatic generation of the input for the new simulation runs. As FEM devices are generated at the beginning of each run, the device and process parameters can be adjusted. The examples of the automated transmission line pulse (TLP) for safe operating area (SOA); conventional, high-voltage, and transient latch-up (TLU) cases; statistical analysis of the process capability index C_{pk}; and system-level ESD IP development cases are summarized next.

11.3.1 SIMULATION AUTOMATION FOR TLP AND PULSED SOA

Although TLP measurements do not correspond to any qualification standard, the method is useful for benchmarking and comparative characterization of the stand-alone ESD clamps and IC pins pulsed performance, for example, with the hardware tools [12,13]. In particular, for an active device with a separate control electrode, TLP can be used to evaluate SOA to determine the protection window and self-protection current capability. Similarly, the dependence of the triggering parameters of ESD devices upon filed electrode bias or injection electrode current can be measured and used for clamps design.

The algorithm of pulsed SOA simulation is similar to the corresponding experimental technique. For MOS (BJT) devices, pulsed drain-source (collector–emitter) I–V characteristics are simulated at the constant gate bias (base current) conditions. The TLP simulation must account for dV/dt coupling and determine electrode current with a certain time delay from the beginning of the pulse. In general, it requires one transient simulation for each I–V data point. Running TLP simulations with traditional tools is possible but is complicated because of the nonlinear device characteristics with significant change of the device impedance on the bias conditions and the necessity to manually extract the current and voltage values from the waveforms at the end of the pulse (e.g., between 70 and 90 ns from the beginning of 100 ns pulse).

An alternative approach can be implemented using automation in the simulation tool that supports parameterization. In this case, the numerical algorithm automatically calculates the amplitude of TLP voltage based on the results extracted from the previous transient run. In practice, the TLP pulse amplitude is determined by taking into account additional parameters governing the current and voltage steps. In addition to the conventional pulse parameters, the algorithm control parameters include the initial voltage step *Vstep*, the voltage step size in the vicinity of the critical points *Vprec*, the current multiplication factor *Imult*, and threshold current *Ithr* (Figure 11.15). The TLP algorithm automatically generates a number of necessary transient simulation runs calculating data points for the entire TLP I–V characteristic. Simulation of each I–V characteristic is started from the minimum TLP source voltage *Vmin* and continues with voltage steps *Vstep* until the critical threshold current level *Ithr* is reached. The TLP source amplitude is automatically adjusted to extract the characteristic points on the I–V curve with accuracy determined by the values of *Vprec* and *Imult* parameters.

For example, in case of NMOS, the entire family of the output drain source I_d-V_{ds} characteristics for different gate bias V_{gs} values is obtained using the preceding algorithm, which guarantees the minimum amount of simulation runs and, consequently, the shortest simulation time. Sufficient number of data points is automatically calculated to resolve both the on- and off-state regions for different control electrode conditions and the negative differential resistance and the high saturation current

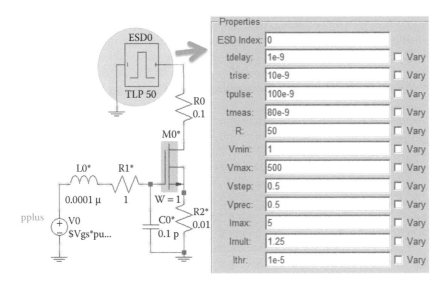

FIGURE 11.15   Mixed-mode circuit and interface for automated TLP.

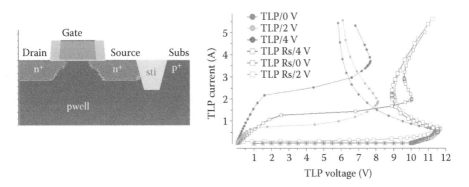

FIGURE 11.16   Cross section and pulsed SOA for NMOS with different source lumped resistor.

regions (Figure 11.16). The pulsed SOA characteristics for different device parameters can be calculated and compared within the same simulation project.

11.3.2   MIXED-MODE SIMULATION AUTOMATION FOR LATCH-UP

The standard latch-up test procedure JES78 [14] can be understood on a simplified level as two types of the electrical pulsed tests—the overvoltage and the injection. To pass the test, neither overvoltage nor injection should result in sustained change (e.g., less than 10%) of the power supply current after the pulse application. The overvoltage test is verification of the ability of the power supply pins of the IC to withstand 50% increase of the power supply voltage over the specified maximum operating voltage (MOV) rating, unless the current limit, for example, 100 mA, is reached first. The injection test is verification that the current injection through the output or input pins does not result in similar increase in the power supply current.

The standard latch-up test is typically defined with the current injection levels either 100 or 200 mA and the pulse length 5–5 ms. The standard latch-up test procedure includes many aspects related to the pin-type definition, pin grouping, and test voltage limitation. Meantime, the physical phenomena in the case of the system-level stress in power-on conditions are somewhat similar to the conventional latch-up while being observed at much higher current levels of system ESD or surge stress in the corresponding time domains. Therefore, the physical simulation problem statements for both latch-up and system-level stress in power-on condition are similar. Upset of the system due to the system-level stress pulse can result in permanent turn-on of the parasitic structures with negative differential resistance observed in a particular IC layout design if the holding voltage of the device structure is below the applied power supply voltage. The permanent high conductivity state can result in burnout of the active Si structure or interconnect regions due to high dissipated power or high current. Thus, the simulation methodologies described for the latch-up are directly applicable for simulation of more complex system-level cases including TLU scenarios. The mixed-mode simulation problem can be stated separately for low-voltage (LV) and high-voltage (HV) latch-up cases. The LV CMOS latch-up is the result of the turn-on of the parasitic pnpn structure with low-holding voltage. The HV latch-up is the result of the turn-on of the parasitic npn structure formed by the n-pocket regions and p-substrate.

For LV CMOS *I/O* and *core* latch-up scenarios, the conditions of a parasitic silicon-controlled rectifier (SCR) structure turn-on can be met by injection, over-voltage, or a single upset event from ionizing radiation source. In *core* latch-up case, the parasitic SCR structure is formed by a pair of high-side (HS) p$^+$-emitter and low-side (LS) n$^+$-emitter regions in close proximity, isolated by corresponding well regions.

For example, HS-connected PMOS and LS-connected NMOS devices in the inverter stage are victim structures potentially susceptible to latch-up. A forward-biased junction in the remote I/O circuit block can become an injection source. A physical equivalent FEM device for the latch-up study is a single multiterminal structure with both core victim and injection I/O structures. The worst-case pnpn structure for the core latch-up has side by side drains of MOS devices isolated by the well regions with contact diffusion at the maximum source to body spacing and the injector p–n junction, for example, I/O diode (Figure 11.17). In corresponding mixed-mode circuit, this structure can reproduce the effects of the injector source isolation. The structure combines the core victim CMOS components in power-on condition and the current injection diode.

In case of I/O latch-up, the parasitic pnpn structure is already present inside the I/O circuit block. During the test the carriers are injected into the structure. For example, in the I/O output buffer, the SCR emitter and bases are formed by the corresponding sources and bodies of the MOS devices (Figure 11.18a). During the latch-up tests, the output is either pulled up above the power supply level (Figure 11.18b) or pulled down below the ground level (Figure 11.18c). It creates conditions corresponding to

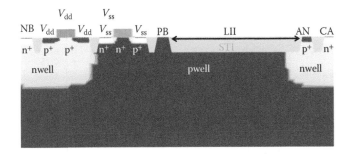

FIGURE 11.17   Equivalent structure for the core latch-up simulation.

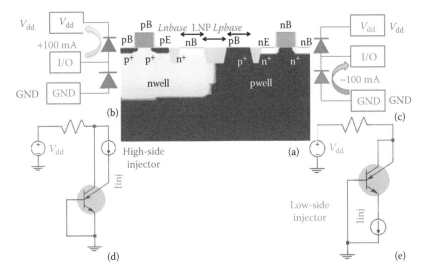

FIGURE 11.18   Device cross section for physical simulation of the I/O latch-up (a), diagrams for the latch-up test for HS (b) and LS (c) injections, and corresponding mixed-mode circuits (d, e).

either the HS hole injection in the PMOS body or the LS electron injection in the NMOS body. The events can be analyzed using the transient simulation analysis with the simple circuits (Figure 11.18d and e) including the FEM device (Figure 11.18a). The I/O buffer simulation can be done to determine the adequate isolation by the body guard rings *Lnbase* and *Lpbase*.

The analysis of FEM structures (Figures 11.17 and 11.18) requires substantial variation of the structure geometry parameters and significant structure depth. Solution of this problem with traditional TCAD approach is rather challenging. Running the calibrated process simulation flow for structures up to 100 μm total length with fine process mesh and accurate diffusion models is impractical, unless a substantial simplification of the flow and device structures is used to mitigate the inevitable convergence issues and enormous simulation resources requirement.

On the contrary, application of the parameterized device templates and process for the latch-up analysis is simple. The solution from previous run is used to obtain and

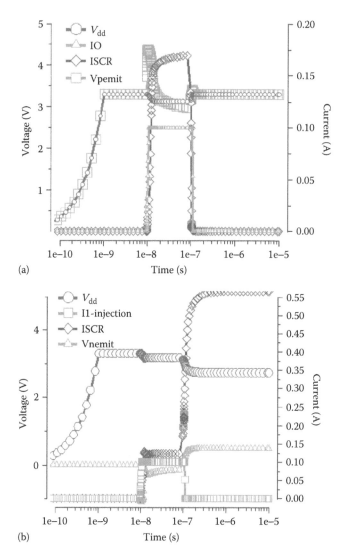

FIGURE 11.19 Conventional pass–fail transient analysis: no latch-up HS (a) and latch-up LS (b) injection cases V_{dd}—the Nbase voltage I0 and I1; the HS injection and the LS injection currents Vpemit and Vnemit—the voltages of the pEmitter and nEmitter ISCR—the power supply current.

import the initial conditions from slow power supply ramp into fast transient simulation runs for latch-up analysis with different structure parameters. In simplified approach, the end-of pulse power supply current *ISCR* value (Figure 11.19) can be placed in the pass–fail table to analyze the layout-dependent latch-up design rules. For example, according to simulation results (Table 11.4), the double guard ring length parameters (Figure 11.18) *Lnbase = Lpbase* > 3 µm guarantee safe passing of 100 mA injection test. The automation can improve on the pass–fail approach. For example, application

TABLE 11.4

Example of Pass/Fail Simulation Outcome

T (K)	LNP (μm)	LNbase (μm)	LPbase (μm)	HS Inject.	LS Inject.
450	0.5	0.25	0.25	LU	LU
450	0.25	2	2	LU	Pass
450	0.25	3	3	Pass	Pass
300	0.5	0.25	0.25	Pass	Pass
300	0.25	2	2	Pass	Pass

problem statements often require an estimate of the critical latch-up current level in already fixed layout. In this case, pass–fail approach is inconvenient.

The automated latch-up simulation algorithm calculates the critical latch-up current as function of the device geometry or process parameters. The algorithm implemented in the latch-up tester (LUT) supports the automatic monitoring of the power supply current level after each injection current pulse (Figure 11.20). Each automatic transient mixed-mode simulation run is similar to the single pass–fail test run (Figure 11.19). The power supply current $Ips1$ before and $Ips2$ after the injection pulse tp are automatically compared at the time point tm. Similarly to the TLP voltage source, the injection current source $Iinj$ level is automatically selected based on the analysis of the results of the previous simulation runs. This way the critical injection current value is determined for latch-up in the structure with given set of parameters.

The LUT circuit element combines the power supply voltage and injection current sources. An example of the core latch-up simulation with multiterminal FEM device (Figure 11.17) and LUT circuit element connected to the core inverter and injecting diode is shown in Figure 11.21a. The resistors R1 and R2 are used to account for the corresponding worst-case scenario body to source separation of the core victim circuit. In the example of numerical analysis for the LV CMOS core latch-up, the critical current for core latch-up is calculated as a function of the injecting diode and core victim device separation distance LII for different initial substrate temperatures (Figure 11.21b).

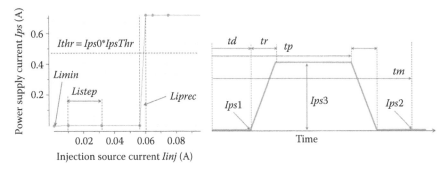

FIGURE 11.20   LUT algorithm for automated latch-up simulation.

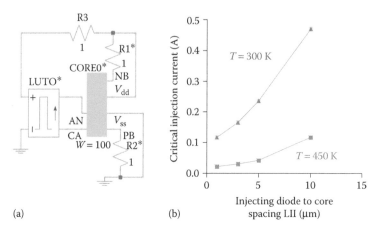

(a)   (b)

FIGURE 11.21 Mixed-mode circuit for automated core latch-up simulation with LUT source (a) and calculated critical current as a function of injector–inverter victim structure separation (b).

Similar principles are extended to more complex HV latch-up cases. In analog design with HV-extended CMOS or BCD processes, the devices from different voltage domains are placed into substrate isolated n-pocket regions. The pockets are formed by corresponding deep N-well or n-Epi with optional N-buried layer regions. Depending on the electrical connection, the two adjacent pockets can form a parasitic npn structure (Figure 11.22). In the latch-up test when the high voltage is applied to the pins connected to the pockets, the internal structures may inject current and create the conditions corresponding to cases of the HS and LS injection according to

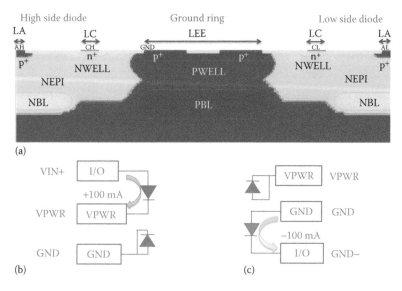

FIGURE 11.22 Structure for HV latch-up analysis (a) and circuit diagrams for the HS hole (b) and LS electron (c) injection.

the diagrams in Figure 11.22b and c. Unlike the case of the conventional LV CMOS latch-up, the main parasitic structure that turns on first is pocket-to-pocket npn. Only then, in the presence of HS p-emitter, the corresponding SCR structure can turn on.

HV n-Epi-to-n-Epi latch-up is a rather complex problem both in practical terms and from the point of view of the physical simulation application. The typical simulation goal is finding critical conditions for latch-up as a function of the physical equivalent latch-up structure parameters representing pocket-to-pocket isolation, regular and active guard rings, Epi region extension, and the physical structure design inside the pocket itself. The HV latch-up analysis requires accounting for the electrothermal effects due to high dissipated power even in current detection regime during the injection pulse. A simplified physical FEM device cross section for multiterminal equivalent structure as shown in Figure 11.22a can be analyzed using the approach described earlier. The example of the HV latch-up simulation with LUT algorithm and HS simulation circuit and extracted dependencies of the critical injection current upon the applied nEpi-nEpi voltage for two LEE nEpi-to-nEpi spacing parameter values is presented in Figure 11.23.

The practical relevance of the approach for the system-level design is the possibility of leveraging the same method for the analysis of a possible unexpected "sneak" current pass in the layout as well as TLU. This method is effective in combination with the layout cutline cross section extraction described earlier. Another system-level problem can be addressed in a similar way. The TLU [15] simulation susceptibility of the on-chip snapback device in particular port-TVS-PCB-IC network can be evaluated as a function of the structure parameters applying similar automation algorithm. System-level ESD pulse-induced TLU scenarios can be studies using the example of the mixed-mode simulation circuit with HMM pulse source and the power supply (Figure 11.24). The stressed pin is connected through the equivalent circuit to the DC power supply voltage source. Depending on the biasing conditions, the dual-direction snapback device (Figure 11.24) can remain in the high conductivity state as a result

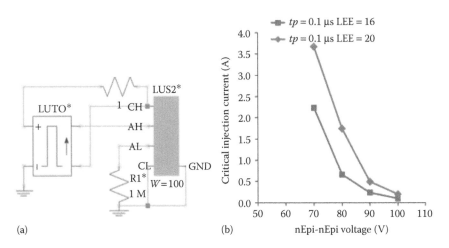

(a) (b)

FIGURE 11.23 HV latch-up simulation circuit (a) for HS injection with LUT algorithm and calculated dependencies (b) for different pocket-to-pocket voltage for two-pocket separation parameters.

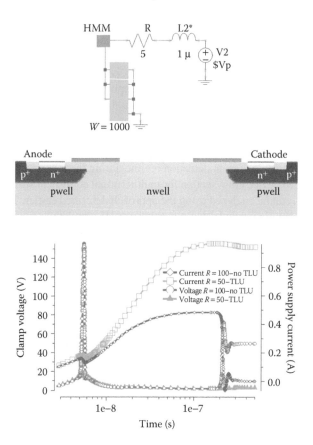

FIGURE 11.24 Mixed-mode analysis for the transient latch-up of stand-alone DIAC device.

of the ESD stress pulse or experience complete turn-off (Figure 11.24). The simple example of TLU as function of the power supply circuit parameters is illustrated by the waveforms (Figure 11.24) for 50 and 100 Ω series resistance values.

11.3.3 AUTOMATED PARAMETERIZED C_{pk} ANALYSIS

An entirely new spectrum of system level-problems solved with parameterized TCAD is the repeatability of the protection window and ESD device characteristics taking into account process variation. Traditionally, the device parameters variation is evaluated based on statistical data accumulated for multiple manufacturing lots using automated electrical tests of the parameters of the scribe line test structures. These statistical data provide for the distribution function for each parameter with the mean μ and standard deviation σ values. Based on the lower and upper limits (*LSL* and *USL*) specified for each parameter, the C_{pk} index is calculated to represent the difference between the mean value and the limits normalized to 3σ: $C_{pk} = \min\left[\left(USL - \mu/3\sigma\right),\left(\mu - LSL/3\sigma\right)\right]$. The C_{pk} serves as an indicator of the tightness of distribution of a given device parameter at given manufacturing facility. It can be used to predict the probability of a device parameter not being within the specification limits based on the assumption of Gaussian distribution for the probability density function $F(\sigma) = \left(1/\sqrt{2\pi}\right)\int_{-\infty}^{\sigma} e^{-t^2/2}dt$ (Table 11.5). In the systems with high reliability requirement for example, the so-called 0 dppm or 6-sigma for automotive or medical applications ESD protection solutions should have $C_{pk} = 2$ or larger. Thus at the same equal conditions, the specification limits for ESD devices in the design with $C_{pk} = 2$ are 2σ higher than in consumer product design margin $C_{pk} = 1.33$ (Table 11.5).

The trend in shrinking of the ESD protection window and design margins in power-optimized processes demands a more careful IP development and control. Meantime, the development learning cycles and test volume for ESD devices is rather low. The simulation-based ESD device design taking into account process variation is difficult to carry out with the standard TCAD tools. Estimate of an ESD solution C_{pk} provides a significant advantage in the design process.

The simulation approach to evaluate the C_{pk} parameters is illustrated on the example of the lateral HV avalanche diode. The high-holding voltage avalanche diodes are one of the key components for system-level cells. The statistical analysis is based on parametric definition of the key structure parameters (Figure 11.25)

TABLE 11.5
Relationship to Process Fallout Measures

C_{pk}	σ Level	Area under $F(\sigma)$	Process Yield (%)	Fallout (DPPM)
1.00	3	0.997300204	99.73	2700
1.33	4	0.999936658	99.99	63
1.67	5	0.999999427	99.9999	1
2.00	6	0.999999998	99.9999998	0.002

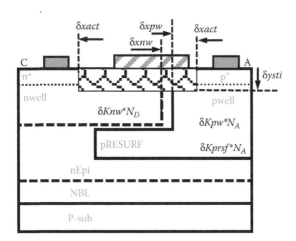

FIGURE 11.25 Defined variation of the lateral HV avalanche diode structure parameters for the automated C_{pk} mixed-mode analysis.

TABLE 11.6

Input Parameters for C_{pk} Simulation

Increment Parameter	Variable	σ
n-Well peak doping	*Knw*	1%
p-Well peak doping	*Kpw*	1%
pRESURF peak doping	*Krs*	1%
nEpi peak doping	*Kep*	1%
STI depth variation	*ysti*	5 nm
Active region misalignment	*xsti*	15 nm
Cathode n-well mask tolerance	*xnwell*	15 nm
Anode p-well mask tolerance	*xpwell*	15 nm
Anode pRESURF mask	*Xprsf*	15 nm

and their variation margins derived from the test data for a given process technology (Table 11.6). The automated generation of the mixed-mode simulation runs is accomplished using the Monte-Carlo algorithm with independent variation of the selected device parameters. Figure 11.26 shows a fragment of the run table with the extracted breakdown voltage at the current level of 10 µA and a plot with entire family of calculated breakdown I–V plots for the lateral avalanche diode.

The methodology can be used to clarify various C_{pk}-related aspects of the specific ESD design solutions. For example, the statistical characteristics of two different ESD device design options can be compared. The example of statistical C_{pk} analysis for double RESURF p–n (Figure 11.27) and compact p-i-n diode

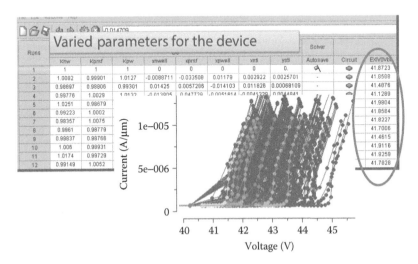

FIGURE 11.26 Automated mixed-mode analysis for C_{pk}: auto-generated mixed-mode run table with extracted breakdown voltage (circled) and a complete set of calculated I–V characteristics.

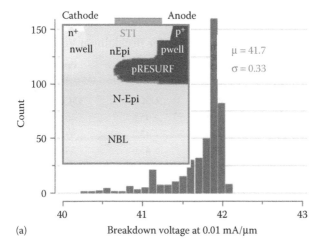

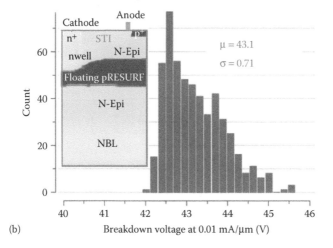

FIGURE 11.27 Device architecture effect on C_{pk} for *Vbr* for the (a) double RESURF and (b) p-i-n lateral HV avalanche diodes automated mixed-mode analysis.

(Figure 11.27) architectures demonstrates different distribution of the breakdown voltage parameter that allows selecting one solution over another based on the capability index requirements. Alternatively, for already finalized design, a reverse problem can be solved by deriving LSL and USL parameters to meet certain C_{pk} level (Table 11.7).

Similarly, the methodology can be used to analyze the C_{pk} for major figures of merit for high-current operation of the ESD devices and clamps. For example, in the case of local protection of the NLDMOS array with the snapback NLDMOS-SCR clamp, the variation of the triggering voltage for both the device and the clamp can be compared to predict the fallout level (Figure 11.28) of the protected pin. The statistical analysis is applied separately to the critical voltage of NLDMOS device

TABLE 11.7

Derived Limits to Meet Different C_{pk} for Breakdown Voltage

Desired C_{pk} for Vbr	Sigma Margins	Double RESURF p–n Diode ($M = 41.7$ V and $\sigma = 0.33$ V)		Pseudo p–i–n Diode ($M = 43.1$ V and $\sigma = 0.71$ V)	
		LSL (V)	USL (V)	LSL (V)	USL (V)
1.33	4σ	40.4	43.0	40.2	45.9
1.67	5σ	40.0	43.3	39.6	46.7
2.00	6σ	39.7	43.6	38.9	47.4

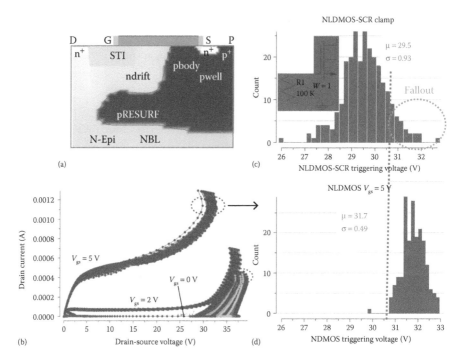

FIGURE 11.28 NLDMOS cross section (a), and the frequency plots for the triggering drain-source voltage at $V_{gs} = 5$ V (b), drain-source I–V characteristics (c), and grounded-gate NLMOS-SCR (d).

output drain source characteristics and to the triggering voltage of the grounded-gate NLDMOS-SCR clamp (Figure 11.28). To improve the fallout, the measure may involve replacement of grounded-gate NLDMOS-SCR with NLDMOS-SCR clamp with HS avalanche diode reference sub-circuit, which is less process variation sensitive. The grounded-gate NLDMOS-SCR clamp is relying in the avalanche injection and high multiplication coefficient of the internal blocking junction in parasitic npn parameters.

This effect is rather sensitive to the blocking junction and gain parameters of the internal npn structure. On the contrary, the clamp with reference sub-circuit

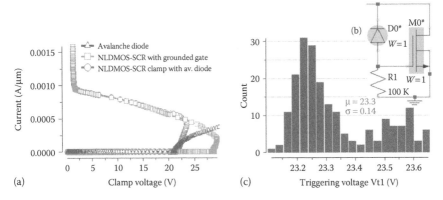

FIGURE 11.29 I–V characteristics (a) of the NLDMOS-SCR clamp with the avalanche diode reference (b) and separate clamp components. Simulated frequency plot for statistical variation of triggering voltage (c) for this clamp.

(Figure 11.29a) has the NLDMOS-SCR triggering under low multiplication coefficient conditions and is less dependent on the npn parameters variation.

Due to the lower process variation sensitivity of the avalanche diode for the triggering characteristics, variability of the clamp (Figure 11.29) has $s = 0.14$ and is much lower than the grounded-gate NLDMOS-SCR clamp (Figure 11.28c) with calculated standard deviation of the triggering voltage parameter $s = 0.93$. This clamp can be used to eliminate the fallout of the case (Figure 11.26).

11.3.4 ESD IP Design with Mixed-Mode Simulation

ESD library is usually released in the form of a small clamp circuits with high-pulsed current capability. Historically, the clamp types are subdivided into so-called active and snapback clamps. Active clamps are designed based on conventional integrated power device arrays to react either to the fast voltage transients or to the voltage above some critical level. In the on-state ESD condition they must provide a current path with positive differential resistance. In active clamps, the clamp state is controlled by a driver circuit implemented as a part of the clamp. Most typical CMOS active clamps are based on a power array with control electrode driven by either a RC-based or an avalanche current reference circuit. Pulsed I–V characteristics of the active clamps are somewhat similar to either a CMOS diode (gate connected to the drain) or an avalanche diode.

An active clamp for system-level protection (Figure 11.30a) can combine both the RC-driver for the fast transient system-level ESD pulse operation and an avalanche diode reference for the slow surge pulse turn-on. The RC driver for the NLDMOS M_1 array (Figure 11.30b) is based on $C_0 R_0 M_0$ components. Under fast-voltage transient conditions before the capacitor C_0 is charged, the M_0 switches to the on-state charging the gate on the array M_1 and thus bringing the active clamp into a low impedance state, which is defined by the saturation characteristics of the M_0. During the slow-voltage rise on the protected node above the avalanche diode D_0 breakdown voltage, the gate of the array M_1 is pulled up. This type of

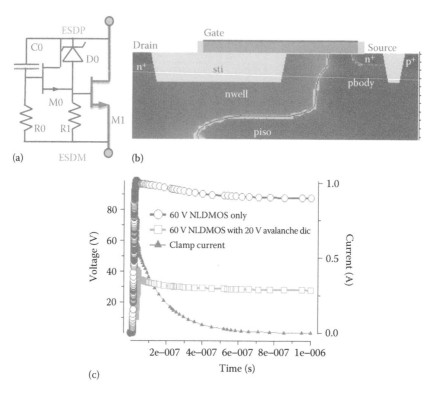

FIGURE 11.30   System-level clamp circuit (a) with combined RC and HS avalanche diode pull-up based on NLDMOS (b) array and the clamp transient waveforms for the system-level ESD and surge pulses calculated as thermal coupled solution with the extracted peak temperature (c).

clamp can be used for protection of pins, which can tolerate high node capacitance, for example, battery charger pins.

Transient mixed-mode analysis taking into account self-heating effects is required to evaluate transient high-current performance of the clamp and corresponding limitations for the ESD (Figure 11.30c), surge pulses, as well as other regimes that may include hot plug-in, power supply, and ground bus ringing and latch-up.

The high-current path in active clamps is provided by integrated components in the monopolar conductivity regime and can often be analyzed using circuit simulation with compact models. However, the regimes involving the system-level pulses require electrothermal analysis, which typically is not supported by the compact models. Thus, physical TCAD analysis is often required. Similarly, the snapback clamps operation is based on conductivity modulation realized in specialized ESD devices. Compact models usually do not cover such nonlinear conductivity modulation regimes, and therefore, evaluation of the snapback clamps operation inevitably requires application of the transient mixed-mode analysis. Due to negative differential resistance, the clamp impedance changes nonlinearly above the critical voltage level. It can also vary with the pulse rise time. Snapback clamps are bi-stable, and their triggering into the low impedance state is load dependent. To suppress current localization in the conductivity modulation state, an additional negative feedback mechanism must be introduced. For example, a ballasting region with current saturation can be used to equalize the current density across the snapback device.

There are only three isothermal conductivity modulation mechanisms [16] observed in practice that can result in negative differential resistance in both the ESD and parasitic latch-up structures (Table 11.8). In spite of all the wide range of snapback ESD clamps and the complexity of specific implementations, the basics range of operating principles is quite limited. The physics of each conductivity modulation mechanism can be studied using simplified quasi-one-dimensional structures, while the more complex 2D FEM devices and clamp can be further analyzed using transient mixed-mode simulation. The negative differential resistance effect is the result of the space charge neutralization by the carriers injected from the forward-biased junctions and generated due to avalanche impact ionization. For example, avalanche injection processes results in negative differential resistance in npn structure, and corresponding S-shaped I–V characteristics can be observed with high load resistance.

A compact device–level dual-direction solution is discussed next to illustrate the system-level simulation analysis specific for the snapback clamps. As two systems might be spatially separated, the ground potentials of the systems can be different. If any of the system ports is protected by unidirectional clamp with body diode, a permanent current flow between the two systems may occur interfering with normal operation. Therefore, dual-direction voltage tolerance is often specified for on-chip ESD protection. Device engineering of the dual-direction solutions presents the biggest challenges for the system ESD protection.

The most straightforward but space-inefficient approach is to construct the clamp from back-to-back (or anti-serial) stack of two substrate-isolated ESD structures. In this case, the middle connection of the blocking junction cathode regions and n-pockets forms a floating node. For example, two lateral avalanche diodes similar to that discussed earlier can be connected forming a dual-directional clamp. However, avalanche diodes provide rather low current capability per micron width, which results in high node capacitance for the system-level currents. Therefore, dual-direction SCR solutions become preferable for signal pins.

Often the dual-direction solution can be obtained by anti-serial substrate-isolated NLDMOS-SCRs. However, a more compact and efficient device-level solution can be achieved with the DIAC structure. It combines the isolated dual-blocking junctions

---

TABLE 11.8
Isothermal Conductivity Modulation Physical Mechanism

Physical Mechanism	Devices	Current (mA/mm)
Avalanche injection	NMOS, PMOS, NPN, PNP	0.1–3
Double avalanche injection	P-i-n, M-i-n diodes	0.1–1
Double injection	LVTSCR, BSCR, LDMOS-SCR, DIAC	10–50

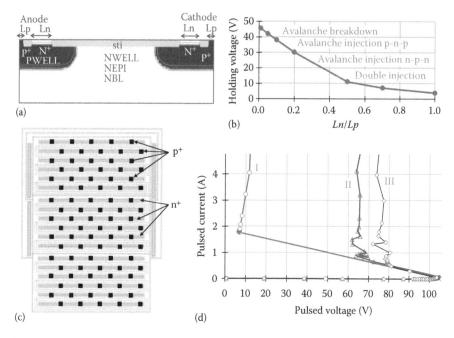

FIGURE 11.31 Simulation: DIAC structure cross section (a) and the dependence of the holding voltage upon n⁺- and p⁺-diffusion length ratio Ln/Lp (b). Experimental: triple-stacked cell with inter-digitated injector placement topology (c) and TLP I–V characteristics for full-width injector regions (I), inter-digitated with p⁺ area 20% (II), and inter-digitated with p⁺ area 10% (III) (d).

with pairs of n⁺ and p⁺ injector regions (Figure 11.31a). The injector regions carry on an interchangeable role acting as bases or emitters depending on the ESD pulse polarity. For each voltage polarity, the corresponding pair of bases and emitters is formed. To adjust the triggering characteristics and holding voltage of DIACs, the n⁺- and p⁺-diffusion length (Figure 11.31b) or area factors in the inter-digitated cell layout (Figure 11.31c) can be varied to control the holding voltage of the DIAC according to the injected carrier balance. This effect can be studied and optimized by the numerical simulation (Figure 11.31a and b) and verified by the experimental results (Figure 11.31c and d) [5].

There are other challenges to the integrated DIAC implementation that require mixed-mode simulation analysis. This includes clamp response to different system-level pulses, vertical and lateral isolation, high-holding voltage requirements, and multi-finger turn-on effects. As on-chip solutions require compact layout capable of sustaining high-current level and latch-up isolated from other internal circuit blocks, it is usually implemented in the form of a compact cell. However, in case of high-voltage structures, the array turn-on can exhibit sensitivity to the pulse type due to nonuniform current sharing between fingers (Figure 11.32).

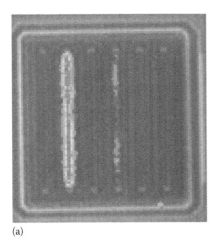

(a)   (b)

FIGURE 11.32 Backside electroluminescence from the DIAC structures in reversible pulse operation with HMM pulse (a) and the de-processed top-level view after air-gap stress burnout (b).

Undesirable non-even multi-finger turn-on effect is often addressed using topological layout approach, which physically implements a continuing long finger approach. Thus, the protection device is implemented as very long single finger, racetrack (Figure 11.33a), horse shoe, or "snake" (Figure 11.33b). In case of fast pulse, all the fingers in the array may turn on simultaneously, but under slow pulse conditions only one finger can turn on and burnout. This may explain miscorrelation between fast contact pulse and slow air-gap pulse test results.

Alternative way of overcoming multi-finger turn-on involves ballasting the ESD current by poly-resistive regions reusing the poly field plates (Figure 11.33c). It can be validated using mixed-mode analysis for the DIAC with embedded poly-ballasting regions demonstrated the change in the current distribution across the multi-finger device cutline (Figure 11.34a through d) and were confirmed experimentally (Figure 11.34e and f) [5].

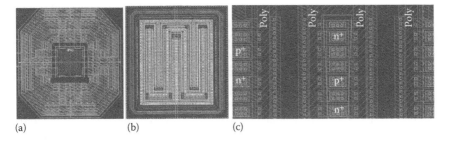

(a)   (b)   (c)

FIGURE 11.33 Methods to reduce the non-even multi-finger turn-on effect in DIAC using "race track" (a), "snake" (b), layout design, and poly resistor ballasting (c).

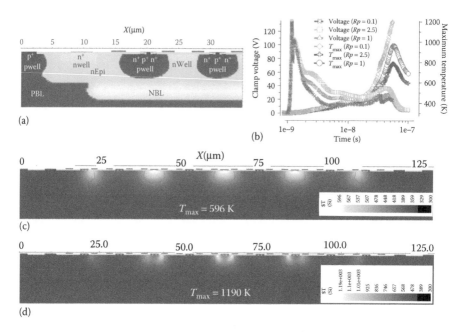

FIGURE 11.34 Fragment of left side of the four-finger DIAC structure with the ballasting poly field plates (a), calculated transient dependencies of the peak temperature for different plate lumped resistance Rp (b) for 8 kV HMM pulse, and the temperature depth profiles for $Rp = 2.5\ \Omega$ (c) and $Rp = 0.1\ \Omega$ (d).

11.4 CONCLUSION

An effective system-level ESD design is essentially similar to the approach adopted in modern physics: *Make a Guess—Calculate the Consequences—Verify Them by Experiment.* However, unlike decades ago the required accuracy can hardly be achieved using analytical methods. For integrated semiconductor devices with complex doping profiles, regions, and nonlinear conductivity modulation effects, only numerical simulation can offer adequate accuracy to reduce the amount of time and resources spent on experiments down to a reasonable limit. Thus numerical simulation becomes the only realistic alternative to help ESD device and circuit engineers in addition to pure empirical approach. Correspondingly, the ESD device and circuit design flow can be transformed as *Make a Hypothesis—Numerically Simulate the Consequences—Verify Them by Experiment.* The TCAD methods utilized properly within the well-understood limits of their physical applicability enable the highly efficient ESD system and chip co-design approach. The requirements for highly integrated SoC/SoP power-efficient analog ICs with high data rates and high reliability requirements can be adequately addressed with accurate physical design using mixed-mode simulation. The proposed and verified approach to achieve this goal includes (1) one-time application of physical process simulation or utilization of the experimental data to extract the doping profile parameters, (2) creation of parameterized device templates for the FEM devices, and (3) the automated mixed-mode circuit simulation that allows for real-time variation of the circuit and device parameters.

REFERENCES

1. Estmark K., Gossner H., Stadler W., *Advanced Simulation Methods for ESD Protection*, Elsevier, 2003.
2. DECIMM™ Angstrom Design Automation, Release 6.0, www.analogesd.com.
3. Vashchenko V.A., Shibkov A.A., *TCAD Methodologies for Industrial ESD Design*, S4, IEW, 2013.
4. Vashchenko V.A., Shibkov A.A., *ESD Design for Analog Circuits*, Springer 2010.
5. Vashchenko V.A., Scholz M., *System Level ESD Protection*, Springer. 2014.
6. "IEC 61000-4-2:2008, Electromagnetic compatibility (EMC)—Part 4-2," 2008.
7. "ISO 10605:2008 Road vehicles—Test methods for electrical disturbances from electrostatic discharge," 2008.
8. "ANSI/ESDA SP5.6-2009—ESD testing—Human Metal Model (HMM)—Component Level," 2009.
9. "IEC 61000-4-5 Electromagnetic compatibility (EMC)—Part 4-5: Testing and measurement techniques—Surge immunity test." http://www.iec.ch/emc/basic_emc/basic_emc_immunity.htm.
10. Powell D.E., and Hesterman B., "Introduction to voltage surge immunity testing." http://www.denverpels.org/Downloads/Denver_PELS_20070918_Hesterman_Voltage_Surge_Immunity.pdf.
11. Marum S., Kemper W., Lin Y.-Y., Barker P., *Characterizing Devices Using the IEC 61000-4-5 Surge Stress*. http://www.riss.kr/search/detail/DetailView.do?p_mat_type=e21c2016a7c3498b&control_no=293d57e7926f26faffe0bdc3ef48d419.
12. HANWA Tester, http://www.hanwa-ei.co.jp/english/seihin_11.html.
13. HPPI Tester, http://www.hppi.de/.
14. JESD-78D—IC Latch-Up Test, JEDEC 2011. http://www.springer.com/us/book/9780387745138.
15. Ming-Dou Ker and Sheng-Fu Hsu, *Transient-Induced Latch-up in CMOS Integrated Circuits*, Wiley, 2009.
16. Vashchenko V.A., Sinkevitch V.F., *Physical Limitations of Semiconductor Devices*, Springer, 2008.

12 ESD Protection of Failsafe and Voltage-Tolerant Signal Pins

David L. Catlett, Jr., Roger A. Cline, and Ponnarith Pok

CONTENTS

12.1 INTRODUCTION

From the circuit designer's point of view, the amount of current during operation that can be tolerated between a signal input/output (I/O) pin and its associated power rail, when the supply is absent or floating, determines a broad class of circuit topologies known as *failsafe* (FS), *pseudo-failsafe* (PFS), or *non-failsafe* (NFS). This spectrum of circuit topologies is shown graphically in the Venn diagram in Figure 12.1 and broadly defines the outline of this chapter, which will clearly define the attributes of each of these circuit topologies, so I/O designers and electrostatic discharge (ESD) developers approach co-design with a common set of definitions

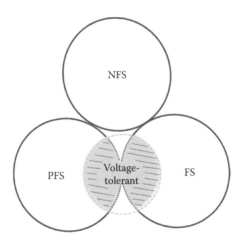

FIGURE 12.1   Venn diagram showing circuit topologies.

and expectations. This is a critical aspect of the chapter, because it is primarily the FS/PFS/NFS topology designation that determines the type of ESD protection that must be applied to the signal I/O pin.

A brief section on NFS circuits and ESD protection will be given primarily to contrast the PFS and FS methods, because NFS ESD protection methods and simulation techniques are already well covered in the literature. PFS applications will also be discussed; however, in many, but not all, low-voltage (LV) complementary metal-oxide semiconductor (CMOS) cases (on which this chapter is focused where typical rail-to-rail swings are 5 V or less), the ESD protection can be developed with methods similar to those used for NFS topologies. Exceptions to this general rule will be noted and may fall closely in line with methods and challenges encountered with FS topology protection. FS topologies, as the title of the chapter suggests, will occupy the most attention as they are typically the most restrictive and most difficult topologies to protect during an ESD event. Last, from an ESD point of view, voltage-tolerant (VTOL) topologies will be discussed, which will primarily include critical circuit integration issues because the ESD protection is largely determined by the current constraints of the I/O (FS/PFS/NFS). This chapter is mainly written for the benefit of circuit designers to gain a better understanding of the challenges that exist when FS or FS/VTOL ESD protection is requested and to provide a checklist of considerations in the upstream development phase for successful integration of their designs. To this end, Table 12.1 concisely summarizes the basic decision tree for the type of ESD protection needed for the broad topology types shown in Figure 12.1. Communicating such concerns early in the design and concept phase, and engaging with ESD developers in far upstream co-design activities, is the least costly pathway for meeting the needs of the application and ultimately the needs of the customer.

TABLE 12.1

Decision Matrix for 3.3 V/5 V ESD Protection Schemes

POWER-OFF Pin-to-Supply Current Tolerance	Nominal I/O Supply Voltage	POWER-ON V_{MAX} at the I/O (V)	I/O Designation	ESD Protection
Zero	3.3	3.3	3.3 V FS	3.3 V FS
Zero	3.3	5	3.3 V FS/ 5 V TOL	5 V FS
Zero	5	5	5 V FS	5 V FS
Limited	3.3	3.3	3.3 V PFS	3.3 V PFS
Limited	3.3	5	3.3 V PFS/ 5 V TOL	5 V PFS
Limited	5	5	5 V PFS	5 V PFS
Not limited	3.3	3.3	3.3 V NFS	3.3 V NFS
Not limited	5	5	5 V NFS	5 V NFS

12.2 DEFINITIONS

12.2.1 NON-FAILSAFE

Mainstream CMOS I/O applications that are defined as NFS signal I/Os, Figure 12.2a, are utilized in applications where the associated power supply is always present (i.e., never 0 V or floating) and the signal is not driven above the power supply by external circuitry or system requirements. If the NFS signal I/O buffer overshoots or is driven externally above its associated power supply, then the signal will be clamped to one forward diode voltage drop above the power supply. If the NFS signal I/O buffer is driven externally while the power supply is absent (i.e., 0 V or floating), then the current will flow into the I/O signal to the extent of compliance provided by the external driver and energize the powered down chip without damage or compromise of the I/O buffer reliability, by maintaining a maximum delta of one forward diode voltage drop to its associated power supply. In such applications, the ESD protection network deployed is often an active rail clamp system that can be simulated and for which numerous studies and tutorials exist [1–3]. Within this context, Table 12.1 lists the "power-off" pin-to-supply current tolerance as "not limited." As NFS topologies and associated ESD protection are well covered in the literature, this chapter will focus on the FS and PFS regions of the Venn diagram in Figure 12.1.

12.2.2 PSEUDO-FAILSAFE

Figure 12.2b shows a typical PFS topology, in which only a finite amount of current is permitted to flow from pin-to-supply when the I/O is driven externally while the power supply is absent (i.e., either 0 V or floating). Under such PFS conditions, the I/O buffer will draw a limited amount of current from the external voltage being

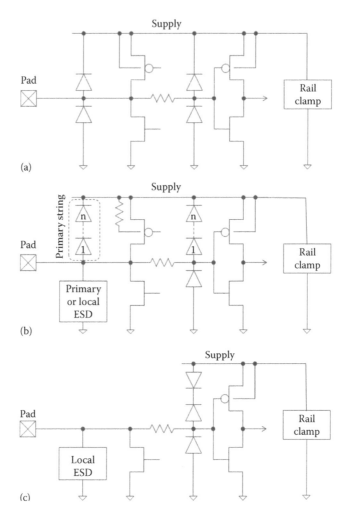

FIGURE 12.2   Typical I/O topologies. (a) NFS I/O, (b) PFS I/O, and (c) FS I/O.

applied, but not enough to energize the powered down chip. Hence, Table 12.1 lists the "power-off" pin-to-supply current tolerance as "limited" for PFS applications.

12.2.3   FAILSAFE

Figure 12.2c shows a typical "open drain" FS topology, which allows for no current path from the I/O to the power supply when the I/O is driven externally while the power supply is absent (i.e., either 0 V or floating). Under such FS conditions, the I/O buffer will draw no current from the external voltage being applied and will not energize the powered down chip. For this reason, Table 12.1 lists the "power-off" pin-to-supply current tolerance as "zero" for FS applications. The FS topology of the I/O dictates the type of ESD protection required, which will be henceforth referred to as a *local clamp*, where *local* denotes a dedicated ESD protection circuit from the specific I/O pin to ground without utilization of the supply rail as part of the ESD shunt path as is available with PFS and NFS topologies.

12.2.4 VOLTAGE-TOLERANT TOPOLOGIES

While the criteria for FS and PFS designations are based on current tolerance when the supply is in the "power-off" condition, VTOL I/Os are defined in the context of FS/VTOL and PFS/VTOL topologies by the capability of the I/O signal to tolerate a higher voltage than the supply voltage when in a "power-on" condition. Combining the definitions of NFS and VTOL is a paradox because NFS signal I/Os are always utilized when the part is powered-on and are not driven externally above the associated power supply. Figures 12.3a and 12.3b show examples of typical PFS/VTOL and FS/VTOL topologies, respectively. True VTOL I/Os, also referred to as *dual-voltage* or *mixed-voltage* I/Os, utilize circuit techniques to ensure that I/O signal voltages in excess of the supply voltage do not damage or compromise the reliability of the I/O buffer over the expected use profile (voltage, temperature, time).

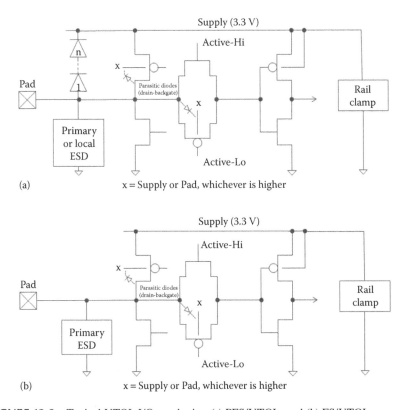

FIGURE 12.3 Typical VTOL I/O topologies. (a) PFS/VTOL; and (b) FS/VTOL.

12.3   CHALLENGES FOR FS SIGNAL PINS

"Open-drain" I/Os, or some topologies where the PMOS well and gate are actively controlled to prevent any conduction via the drain diode path or channel, are typical for FS I/O pins. FS pins on CORE (LV) logic may be protected by a series of diodes, where the number reflects the leakage tolerance of the pins. For higher pad voltages, it becomes necessary to use a parasitic-based ESD cell to implement FS ESD protection to mitigate excessive voltage and leakage at the pad. Diode strings will mitigate the excessive voltage but increase leakage as the number of diodes in series becomes longer. As the components used in the ESD protection are generally the same components deployed in the circuitry being protected, a triggering mechanism for the parasitic ESD pathway needs to be employed to avoid significant circuit leakage or even breakdown prior to the ESD protection engaging. Behavior of these parasitic elements can vary from fab-to-fab if there are no independent masking layers used to determine their construction. TCAD can be used [4–6] as a guideline for understanding these parasitic dependencies, but thorough Si characterization is generally required for FS protection in advance of product deployment.

Local protection clamps that provide direct signal-to-ground pathways without utilizing the power rail come in several broad varieties of so-called snapback devices. These ESD protection devices rely on low-resistance, edge- or level-triggered, parasitic bipolar transistor conduction paths (NPN, PNP) or silicon controlled rectifier (SCR) action to shunt ESD current to ground. Types of typical snapback devices are given in Table 12.2. Use of such devices require careful integration and characterization of the ESD protection before deployment because, generally speaking, snapback devices are not simulated in the way that NFS (active rail clamp system) protection networks are done.

As technology scales and the cost pressure for reduced area and high performance are considered, upstream co-design activities between the I/O design team and the ESD design team are important for integrating the ESD protection into the circuit topology and layout landscape (i.e., proximity and bus resistance considerations). Several focus areas for these integration co-design activities are discussed in this section from the physics perspective relative to the design measures that may need to be considered.

TABLE 12.2
Typical Snapback ESD Devices

Clamp Topologies	Clamp Types	References
Bipolar	GG-NMOS (grounded gate NMOS)	[7–9]
	NTNMOS (NMOS triggered NMOS)	[4,10–12]
	GG_PTNMOS (grounded gate PNP-triggered NMOS)	[10,11,13]
SCR	GG-DINMOS (grounded gate diode-isolated NMOS)	[14–16]
	DTSCR (diode-triggered SCR)	[17–19]
	LVTSCR (low-voltage triggering SCR)	[20–22]
	PNP-triggered SCR	[23,24]
	SBSCR (self-aligned STI-blocked SCR)	[25]

12.3.1 PROXIMITY EFFECTS

As CMOS technologies have moved to non-EPI substrates and associated higher substrate resistances for cost-effective processing, parasitic device interactions have become more important to consider, particularly as technology scales to smaller dimensions. The physics of proximity effects on parasitic elements of the ESD protection need to be well understood and well characterized in advance of upstream co-design activity for FS buffers. In many CMOS applications, the chip substrate plays an integral role in triggering the FS ESD protection, thus control of the interactions with nearby diffusions is critical to success. For example, NTNMOS (*NMOS Triggered NMOS*) protection topology has been well characterized to this end. Of particular importance for the NTNMOS topology is the presence of nearby p-type and n-type diffusions tied to ground [10]. Figure 12.4 shows a simplified representation of this protection topology, which relies on efficient pumping from the source of the pump NMOS transistor to the local substrate (not tied directly to ground) of the clamp NMOS transistor to produce a parasitic NPN to shunt the ESD current from I/O to ground. An n-type diffusion to ground placed in close proximity to the pump NMOS acts as a competitive NPN pathway for the charge intended to pump the local substrate of the clamp NMOS. Consequently, the capacity for shunting the ESD current is diminished. In addition, any p-type diffusion to ground placed in close proximity to the clamp NMOS transistor will diminish local substrate pumping, which will in turn diminish the capacity to shunt ESD current. This type of ESD protection has been well characterized both in Si and through TCAD simulation [4], leading to guidelines for integration of this ESD topology.

While SCR-based protection offers the advantages of a smaller layout footprint and large current-carrying capability, there are also layout integration challenges. SCRs can be intentionally triggered during an ESD event by external circuit elements that will either elevate the PWELL (NPN triggering) or pull down the NWELL (PNP triggering) to reduce the natural triggering voltage for acceptable use

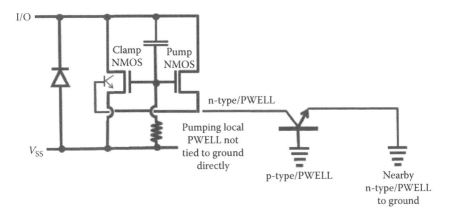

FIGURE 12.4 Parasitic NPN elements in proximity to the NTNMOS ESD protection topology.

in protecting lower-voltage components. On the other hand, they are also susceptible to charge injection from parasitic elements that may trigger the same effects in both wells to activate the SCR outside of the intended triggering mechanism. In normal protective measures for latch-up (LU) immunity, wells around parasitic SCR elements are tapped to prevent injection current from de-biasing the well potentials and turning on the SCR. However, for ESD protection that deploys an SCR as its clamp, tapping the wells will be detrimental to intentionally triggering the SCR during an ESD event. Immunity of the SCR from unintentional triggering, then, cannot come from tapping the wells but must instead be accomplished by preventing the injected charge from getting to the wells (collectors) of the SCR. As with the earlier discussion for the NTNMOS proximity effects, a clear understanding of the parasitic interaction physics that may unintentionally trigger the SCR is important when looking at complex circuit topologies and determining the appropriate measures required for the SCR to trigger predictably. The parasitic elements that may affect the operation of the SCR will be acting as injection sources using the well of the SCR as collectors, or as unintended parallel resistance paths interfering with SCR triggering. From this basic point of view, the immunity schemes that can be considered (1) define an exclusion zone around the SCR to keep problematic parasitic injectors at a safe distance and (2) provide guard rings around the SCR and injection sources to collect stray charge and prevent interaction with the SCR. In general, a combination of both techniques is appropriate, depending on the nature of the parasitic interaction as well as I/O area and cost considerations.

As shown in Figure 12.5a, nearby diffusions to ground (i.e., n-type/PWELL, n-type/NWELL, or p-type/PWELL) are important to consider for SCR operation. N-type diffusions to ground have little impact on the SCR performance if they have small effective area, like an n-type guard-ring surrounding the SCR. N-type diffusions can create a parasitic NPN element (NWELL-PWELL-n-type), which can

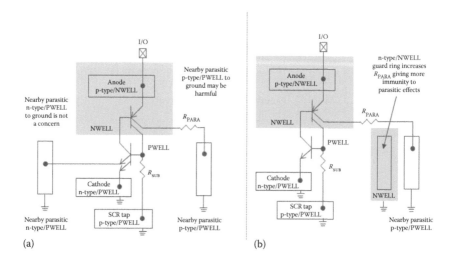

FIGURE 12.5 Parasitic paths effecting SCR performance (a) Nearby parasitic p-type/PWELL and (b) Resistive barrier.

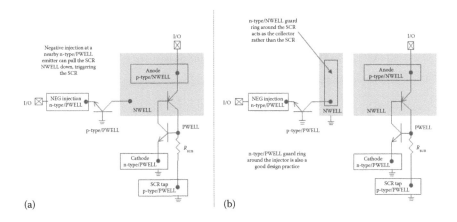

FIGURE 12.6 I/O to I/O parasitic paths effecting SCR performance (a) Negative injection n-type/PWELL and (b) Preferred n-type/NWELL collector.

provide an alternative, but not necessarily a competitive path with the intended SCR path to the cathode. Conversely, n-type diffusions tied to an I/O pin in close proximity to the SCR may act as a negative injection source, as shown in Figure 12.6a. Negative injection can pull down the SCR NWELL, acting as a collector, forward biasing the SCR anode and triggering the SCR. The standard method for collecting negative injection charge is an n-type guard ring (n-type/NWELL) surrounding the injector, tied to ground or power, as shown in Figure 12.6b. While tying the n-type guard ring to ground reduces its efficiency as a bipolar collector, it is less susceptible to act as a parasitic ESD path than when it might be tied to power. The n-type guard ring will act as the preferred collector for the negative injection current instead of the SCR NWELL, while posing no threat to SCR operation.

More problematic are p-type parasitic elements tied to ground in close proximity to the SCR, as shown in Figure 12.5a. The triggering point of the SCR is strongly dependent on R_{SUB}, which is generally controlled by the position of the SCR p-type tap by design. Any lowering of the intended R_{SUB} will make the device harder to trigger, increasing V_{TRIG}, as illustrated in Figure 12.7. The proximity of the p-type element tied to ground offers a parallel resistance path to the intended R_{SUB} for the SCR, thus reducing the effective R_{SUB} and leading to increased V_{TRIG}. Note that once the SCR has triggered, no degradation in current-carrying capability is seen due to the proximity of the p-type element. Increasing the parallel resistance path to these parasitic p-type diffusions to ground can be done by defining an exclusion zone inside of which no p-type diffusions to ground may be placed. This strategy is hard to define *a priori* and will depend strongly on the resistivity of the wells and substrate in definition of the effective distance over which the parasitic element acts. An alternative to a large exclusion zone is the deployment of an n-type guard ring surrounding the SCR, as shown in Figure 12.5b. As discussed earlier, n-type diffusions tied to ground do not pose a risk to the SCR triggering or operating. In this particular case, the n-type guard ring serves the purpose of increasing the parasitic resistance, thus better isolating

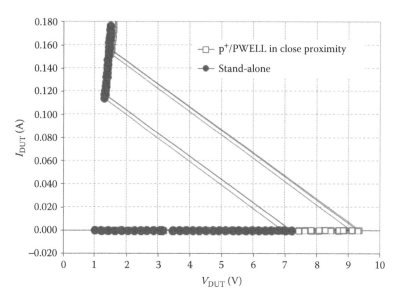

FIGURE 12.7 Parasitic effects on SCR trigger voltage.

the effective R_{SUB} for the SCR. The n-type guard ring is acting as both a collector and a resistive barrier.

P-type parasitic elements, when the emitter is connected to supply or another signal, provide pathways for positive injected charge to tamper with the well potentials of the SCR, bypassing the intended triggering mechanism and creating unintended current draw. Figure 12.8 shows the case of a positive injection source (p-type/NWELL tied to power). Positive injection current, if uncollected, will easily elevate the substrate (PWELL) of the SCR, turning on the lateral NPN and triggering the SCR. Because of the high sensitivity of the SCR to parallel resistance paths to nearby p-type diffusions to ground, surrounding the SCR with a p-type guard ring tied to ground, which

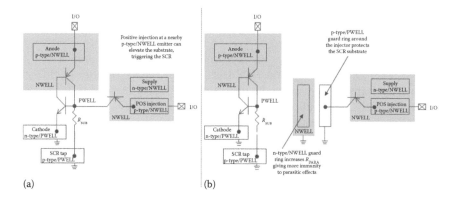

FIGURE 12.8 I/O to I/O parasitic paths effecting SCR performance (a) Positive injection p-type/NWELL and (b) Preferred p-type/PWELL collector.

would act as a preferred collector of the positive charge, is not the right approach. In this case, the best approach is to move the positive injection source as far away from the SCR as allowed by design and area constraints and to collect the positive injection charge close to the injection source by a p-type guard ring, as shown in Figure 12.8b.

12.3.2 NMOS Drain Ballasting for FS Outputs

The obvious way to protect the drains of a FS output buffer from damage during the high-voltage snapback event of the ESD trigger is to make sure that the triggering voltage of the ESD device is below the breakdown voltage of the drains in the buffer. Depending on the FS ESD protection scheme deployed, trigger voltage variation may not be easily done. In the case where the triggering voltage of the ESD protection is close to the breakdown voltages or onset of conduction of the drains in the buffer, techniques need to be deployed to increase the V_{FAIL} of that buffer components to a safe voltage above V_{CLAMP} of the ESD protection.

A common way to design the output buffer for better ESD immunity is to introduce a ballasting resistance in the drain of the buffer of sufficient size to produce enough IR drop at the I_{FAIL} condition of the buffer component so as to increase V_{FAIL} beyond V_{CLAMP}. The cost of this measure is drive strength reduction, which may need to be compensated by increased buffer size. In typical FS output buffer topologies, the weakest link is the NMOS transistor drain, which may go into NPN snapback during an ESD event. I_{FAIL} for this path generally correlates to the width (W) of the NMOS transistor. This may be a little more complicated with layouts where multiple finger transistors are deployed and "multiple-finger parasitic NPN turn on" was not considered in the design parameters. In such a case, the safe approach is to consider the I_{FAIL} of a single finger, rather than the total transistor, of the pad-connected NMOS in deciding the appropriate resistance for ballasting.

Such an approach can be quite costly, and for the case of FS buffers that are also VTOL, this may be impossible. Cascoding the NMOS transistor is an alternative way to add ESD robustness to NMOS output transistors [26] so as to increase the V_{FAIL} for the buffer.

12.3.3 Secondary Protection for FS Inputs

Secondary protection of the gates in an input buffer is shown to be common among the NFS, PFS, and FS topologies shown in Figure 12.2. During an ESD event at the input signal pad, the two most difficult scenarios are (1) a positive strike from pad to ground and (2) a negative strike from pad to supply, both shown in Figure 12.9. Consider first the negative strike on the pad relative to supply. The resistor in series limits the voltage at the gate, while the back-to-back diode keeps the voltage clamped from gate to source on the PMOS transistor. In the second case, a positive ESD strike on pad to ground, the resistor in series limits the voltage at the gate, and the NMOS gate to source voltage is clamped by the reverse bias breakdown of the diode. As technology scales to more advanced technology nodes, this scheme becomes problematic when voltage drop across the secondary clamping path exceeds the breakdown voltage of either the NMOS gate to source or the PMOS gate to source, shown

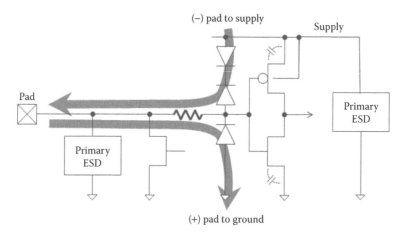

FIGURE 12.9   Problematic ESD pathways for FS input buffers.

in Figure 12.9 as parasitic capacitive elements. Determination of this cross-over voltage depends on comprehensive component characterization early in the development cycle and the ability to model these secondary protection paths, even empirically. While progress has been made for secondary protection on NFS inputs with parallel diode string methodologies [27], this remains a gap for FS inputs as technology scales and should be a factor to consider in designing FS input topologies.

12.3.4   *DV/DT* T RIGGERING ON A S IGNAL P IN

Bipolar-based FS ESD circuits are usually *dV/dt* based because the inherent breakdown voltage of junctions within the same technology node is too high for ESD purposes. A level triggered scheme that may utilize a string of diodes is generally not practical because the ESD trigger voltage may be compromised to mitigate the operational leakage exhibited at elevated temperatures. For low-voltage pins, it may be possible to find a viable design space, but this is not the case for high-voltage pins. As *dV/dt*-based triggering is used at the pin level, care must be taken to track the leakage exhibited at the desired buffer slew rates. I/O designers should characterize buffer performance with the FS ESD cell in place and monitor the leakage exhibited by the ESD components themselves. Again, the NTNMOS example offers a good framework of such characterization and methodology [10]. In this case, the resistor of the RC gate-trigger node can influence the maximum tolerable capacitance at the I/O pin, leading to detrimental coupling during normal operation. Additional circuitry was added to modulate that resistance during normal operation, while still maintaining the effectiveness to protect for short transient ESD events when the supply is floating.

12.3.5   VSS B US R ESISTANCE

Although this section may seem to be obvious, for completeness it must be mentioned that the successful FS ESD protection scheme depends on a low resistance

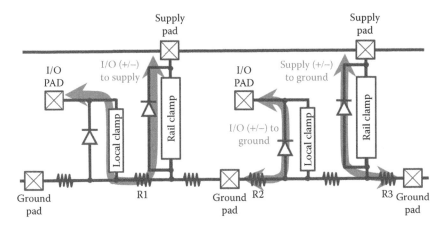

FIGURE 12.10   ESD current paths and associated parasitic bus resistance.

path from the signal pin to ground. As the supply bus is not utilized, only the ground bus resistance needs to be considered. Figure 12.10 shows the generalized ESD protection network and highlights the important paths where low resistance along the ground bus is critical: R1 (resistance of the local clamp to the rail clamp), R2 (local clamp to ground), and R3 (rail clamp to ground).

12.4   CHALLENGES FOR PFS PINS

In PFS applications, only a finite amount of current is allowed to flow when the PFS I/O is driven externally while the power supply is absent (either floating or at 0 V). Under such PFS conditions, the I/O buffer will draw a limited amount of current from the external supply, but not enough to energize the powered-down chip. Referring to Figure 12.2b, current is limited by a string of diodes from I/O to supply. The primary path for ESD current during a positive I/O to ground strike will go through this diode string, drop through the supply bus resistance, and then shunt to ground through the primary supply ESD clamp. This poses a problem from an ESD protection point of view because of the large voltage buildup. This situation can be simulated in the same way we would simulate an NFS network. The primary I/O ESD clamp to ground may be a simple diode from ground to I/O.

If the voltage drop across the diode string is large, a local FS or FS-like protection circuit can be deployed locally as primary ESD protection from I/O to ground that is designed to trigger at a voltage lower than the PFS network. In such a case, the same challenges that we outlined in the FS section apply equally to the local PFS protection from I/O to ground. An additional challenge in this scenario is the possibility of LU or signal latch-up (sLU) involving the high impedance NWELL node on diode 1 (of the primary n-diode string to supply) and the local ESD protection shown in Figure 12.2b. This is particularly an issue if diode 1 is part of the parasitic triggering mechanism of the local ESD protection [11,28]. sLU will not be detected by standard LU testing where only the supply current is monitored. Negative current injection on a signal pin while a neighboring signal pin is high may turn on the vertical PNP of the high impedance NWELL with anode tied to the signal, resulting in unintentional triggering of the local ESD protection. Because of the proximity effects noted earlier for local ESD protection utilizing parasitic elements, care must be taken to balance the effectiveness of the triggering during an ESD event while not increasing LU or sLU risk. In the case cited earlier [28], significant changes to the diode string layout and to the local ESD protection were required to balance these effects.

12.5   CHALLENGES FOR VOLTAGE TOLERANCE

While the current tolerance from I/O to supply determines the broad categories of FS, PFS, and NFS I/O types and ESD protection schemes, a subset of these categories is determined by the maximum voltage seen at the I/O in a power-on condition. If this maximum voltage is higher than the supply voltage for the I/O, then the I/O is considered to be VTOL to the higher voltage. In this case, all components in the I/O and ESD protection must meet reliability requirements at the maximum voltage seen at the I/O. VTOL I/Os are often "dual-voltage" or "mixed-voltage" I/O, deploying switching circuitry to actively regulate the PMOS backgates [29,30], as shown in Figures 12.3a and 12.3b.

Combining the definitions of FS and VTOL, CMOS I/O design techniques must ensure that all pull-down NMOS transistors and all pull-up PMOS transistors tied to the I/O signal pin must be either non-cascoded or cascoded depending on the chosen device voltage ratings. Furthermore, all pull-up PMOS transistors must have both the backgate and transistor gate actively controlled so as to prevent current flow through the PMOS drain-backgate diode or CMOS channel conduction, respectively. With the absence of power supply voltage (i.e., 0 V or floating) and during external voltage being applied to the FS/VTOL signal I/O pin, the I/O buffer will not be damaged or suffer compromised reliability by elevated external DC or transient voltages, will draw no current from the external voltage being applied, and will not energize the powered down chip.

When 5 V-rated NMOS and PMOS transistors are used, the pull-down NMOS transistors and the pull-up PMOS transistors tied to the I/O signal pin can be non-cascoded to meet reliability limits over process (P), voltage (V), and temperature (T) for the respective NMOS and PMOS source, drain, and gate terminals. This is important with the absence of power supply voltage (i.e., 0 V or floating) while the external voltage being applied to the I/O signal pin is 5 V. Otherwise, use of 3.3 V or lower-rated voltage NMOS and PMOS devices must be cascoded to meet the given transistor reliability limits over PVT for the respective cascoded source, drain, and gate terminals. While cascoding satisfies the given NMOS and PMOS transistor reliability requirements, it also creates additional design trade-offs such as body effect, conduction resistance, large layout area consumption, and increased pin capacitance.

In addition to NMOS and PMOS transistor source, drain, and gate terminal reliability, the PMOS transistors tied to the I/O signal pin must have their respective backgates carefully considered in a FS/VTOL application. To meet transistor reliability, avoid drain-backgate diode leakage, and minimize the body effect, the PMOS backgates cannot be simply tied to the given 3.3 V power supply. There must be circuit techniques employed to ensure that the PMOS backgates are driven to either the I/O signal pin voltage (i.e., 5 V) or the given power supply voltage (i.e., 3.3 V), whichever is highest at any time. The transition point between 3.3 and 5 V (and vice versa) should be designed to happen as quickly as possible to minimize any "dead zone," otherwise the PMOS backgates will be floating within the given transition time. If the transition "dead zone" is not minimized, the parasitic PNPs associated with the PMOS transistors will conduct current into the p-type substrate. Depending on the magnitude of the resulting substrate current, it can be manifested in I/O signal pin leakage or power supply leakage or in the worst case be the trigger source of LU in either the I/O circuitry itself or the associated primary ESD protection structure.

As the PMOS transistors tied to the I/O signal pin must have their backgates driven, the NWELL bulk represents a high impedance region that can be modulated because it is not tied directly to a low-impedance power supply. Modulation of the NWELL bulk can be caused by the "dead zone" in the backgate switching circuitry leaving the NWELL floating for a period of time or modulation of the NWELL bulk can be caused by excessive I/O signal undershoot that excites parasitic NPNs that can pull down on nearby high-impedance NWELL regions. In the event the NWELL bulk is modulated below the associated PMOS by at least one Vbe, the parasitic PNP associated with the PMOS transistor will conduct current into the p-type substrate. Depending on the magnitude of the substrate current and proximity to adjacent parasitic SCRs or the primary ESD protection structure, the risk of LU is greatly increased. To mitigate the LU risk, the PMOS backgate switching circuitry must transition quickly between 3.3 and 5 V (and vice versa) to minimize any "dead zone" or floating NWELL scenarios. In addition, the PMOS backgate switching circuitry and the PMOS transistors in the high impedance–driven NWELL regions must be placed as far away from the primary ESD protection structure as possible.

12.6   CONCLUDING COMMENTS

ESD protection for FS and FS/VTOL circuits remains one of the most challenging tasks for I/O design teams and ESD development teams. The first step for this co-design activity is to make sure that I/O designers and ESD development engineers are working off the same set of topology definitions that are provided in Table 12.1. This is an important takeaway from this chapter. Successful integration of such ESD protection requires addressing a list of integration challenges early in the development cycle along with considerations of I/O performance and cost. Among the checklist items discussed were:

- Component characterization early in the technology development cycle to determine the robustness of all components used in both the ESD protection circuitry and the I/O circuitry.
- Understanding the critical interactions and limitations of parasitic elements in the I/O layout that are in proximity to ESD protection schemes that rely on parasitic triggering mechanisms and/or parasitic clamping paths.
- Characterization of the integrated ESD protection scheme with regard to normal operation conditions and reliability.

ACKNOWLEDGMENTS

The authors thank their colleagues Dr. Tim A. Rost and Dr. Vijay Reddy for their careful review and helpful remarks. The authors also thank Dr. Juin J. Liou for his invitation to contribute a chapter to this book.

REFERENCES

1. M.G. Khazhinsky, J.W. Miller, M. Stockinger, and J.C. Weldon, "Engineering Single NMOS and PMOS Output Buffers for Maximum Failure Voltage in Advanced CMOS Technologies," *EOS/ESD Symposium Proceedings*, 2004, pp. 255–264.
2. C.A. Torres, J.W. Miller, M. Stockinger, M.D. Akers, M.G. Khazhinsky, and J.C. Weldon, "Modular, Portable, and Easily Simulated ESD Protection Networks for Advanced CMOS Technologies," *EOS/ESD Symposium Proceedings*, 2001, pp. 82–95.
3. S. Venkataraman, C. Torres, D. Catlett, T. Rost, C. Barr, and K. Burgess, "Design Considerations to Reduce Process Sensitivity for Transient-Triggered Active Rail Clamps in Advanced CMOS Technologies," *Proceedings of the 6th International ESD Workshop*, 2012, p. A.6.
4. G. Boselli, "On the Interaction between Substrate-Triggered ESD Protection Circuits and Embedded Parasitic Elements," *Proceedings of the 4th International ESD Workshop*, 2010, p. A.2.
5. A.A. Salman, G. Boselli, H. Kunz, and J. Brodsky, "Solutions to Mitigate Parasitic NPN Bipolar Action in High Voltage Analog Technologies," *EOS/ESD Symposium Proceedings*, 2010, pp. 309–315.
6. J. Bourgeat, C. Entringer, P. Galy, F. Jezequel, and M. Balfleur, "TCAD Study of the Impact of Trigger Element and Topology on Silicon Controlled Rectifier Turn-On Behavior," *EOS/ESD Symposium Proceedings*, 2010, pp. 11–20.
7. S. Dong, X. Du, Y. Han, M. Huo, Q. Cui, and D. Huang, "Analysis of 65 nm Technology Grounded-Gate NMOS for On-Chip ESD Protection Applications," *Electronics Letters*, 44(19), 2008, 1129–1130.
8. J.-H. Lee, Y.-H. Wu, C.-H. Tang, T.-C. Peng, S.-H. Chen, and A. Oates, "A Simple and Useful Layout Scheme to Achieve Uniform Current Distribution for Multi-Finger Silicided Grounded-Gate NMOS," *Proceedings of the 45th Annual Reliability Physics Symposium*, 2007, pp. 588–589.
9. M.X. Huo, K.B. Ding, Y. Han, S.R. Dong, X.Y. Du, D.H. Huang, and B. Song, "Effects of Process Variation on Turn-On Voltages of a Multi-Finger Gate-Coupled NMOS ESD Protection Device," *Proceedings of the 16th IEEE International Symposium on the Physical and Failure Analysis of Integrated Circuits*, 2009, pp. 832–836.
10. C. Duvvury, S. Ramaswamy, A. Amerasekera, R.A. Cline, B.H. Andresen, and V. Gupta. "Substrate Pump NMOS for ESD Protection Application." *EOS/ESD Symposium Proceedings*, 2000, pp. 7–17.
11. A. Jahanzeb, C. Duvvury, R. Cline, S. Sterrantino, S. Kothamasu, and A. Somayaji, "High Voltage ESD Protection Strategies for USB and PCI Applications for 180 nm/130 nm/90 nm CMOS Technologies," *EOS/ESD Symposium Proceedings*, 2000, pp. 222–230.
12. J. (Heng-Chih) Lin, C. Duvvury, B. Haroun, I. Oguzman, and A. Somayaji, "A Fail-Safe ESD Protection Circuit with 230 fF Linear Capacitance for High-Speed/High-Precision 0.18,urn CMOS 1/0 Application," *International Electron Devices Meeting*, 2002, pp. 349–352.
13. C. Salling, J. Hu, J. Wu, C. Duvvury, R. Cline, and R. Pok, "Development of Substrate-Pumped nMOS Protection for a 0.13 μm Technology," *EOS/ESD Symposium Proceedings*, 2001, pp. 190–202.

14. G. Boselli, V. Vassilev, and C. Duvvury, "Drain Extended NMOS High Current Behavior and ESD Protection Strategy for HV Applications in sub-100 nm CMOS Technologies," *Proceedings of the 45th Annual Reliability Physics Symposium*, 2007, pp. 342–347.

15. Y.-Y. Lin, C. Duvvury, A. Jahanzeb, and V. Vassilev, "Diode Isolation Concept for Low Voltage and High Voltage Protection Applications," *EOS/ESD Symposium Proceedings*, 2009, pp. 1–7.

16. Y.-Y. Lin, V. Vassilev, and C. Duvvury, "The Optimization of Diode Isolated Grounded Gate NMOS for ESD Protection in Advanced CMOS Technologies," *Proceedings of the 6th International ESD Workshop*, 2009, p. E.3.

17. M.P.J. Mergens, C.C. Russ, K.G. Verhaege, J. Armer, P.C. Jozwiak, R. Mohn, B. Keppens, and C.S. Trinh, "Diode-Triggered SCR (DTSCR) for RF-ESD Protection of BiCMOS SiGe HBTs and CMOS Ultra-Thin Gate Oxides," *International Electron Devices Meeting*, 2003, p. 21.3.

18. J. Di Sarro, K. Chatty, R. Gauthier, and E. Rosenbaum, "Evaluation of SCR-Based ESD Protection in 90 nm and 65 nm CMOS Technologies," *Proceedings of the 45th Annual Reliability Physics Symposium*, 2007, pp. 348–357.

19. J. Bourgeat, C. Entringer, P. Galy, P. Fonteneau, and M. Balfleur, "Local ESD Protection Structure Based on Silicon Controlled Rectifier Achieving Very Low Overshoot Voltage," *EOS/ESD Symposium Proceedings*, 2009, pp. 72–85.

20. M.-D. Ker, and H.-H. Chang, "How to Safely Apply the LVTSCR for CMOS Whole-Chip ESD Protection without Being Accidentally Triggered On," *EOS/ESD Symposium Proceedings*, 1998, pp. 1–8.

21. V.A. Vashchenko, M. Concannon, M. ter Beek, and P. Hopper, "High Holding Voltage Cascoded LVTSCR Structures for 5.5 V Tolerant ESD Protection Clamps," *IEEE Transactions on Device and Materials Reliability*, 4(2), 2004, 273–279.

22. Y. Shan, J. He, B. Hu, J. Liu, and W. Huang, "NLDD/PHALO-Assisted Low-Trigger SCR for High-Voltage-Tolerant ESD Protection without Using Extra Masks," *IEEE Electron Device Letters*, 30(7), 2009, 778–780.

23. Y. Morishita, "New ESD Protection Circuits Based on PNP Triggering SCR for Advanced CMOS Device Applications," *EOS/ESD Symposium Proceedings*, 2002, pp. 6–9.

24. Y. Morishita, "A PNP-Triggered SCR with Improved Trigger Techniques for High-Speed I/O ESD Protection in Deep Sub-Micron CMOS LSIs," *EOS/ESD Symposium Proceedings*, 2005, pp. 1–7.

25. K. Kunz, C. Duvvury, and H. Shichijo, "5 V Tolerant Fail-Safe ESD Solutions for 0.18 μm Logic CMOS Process," *EOS/ESD Symposium Proceedings*, 2001, pp. 12–21.

26. W. Anderson, and D.B. Krakauer, "ESD Protection for Mixed-Voltage I/O Using NMOS Transistors Stacked in a Cascode Configuration," *EOS/ESD Symposium Proceedings*, 1998, pp. 54–62.

27. E. Worely, R. Jalilizeinali, S. Dundigal, E. Siansuri, T. Chang, V. Mohan, and X. Zhan, "CDM Effect on a 65 nm SOC LNA," *EOS/ESD Symposium Proceedings*, 2010, pp. 381–388.

28. J. Salcedo-Suner, R. Cline, C. Duvvury, A. Cadena-Hernandez, L. Ting, and J. Schichl, "A New I/O Signal Latchup Phenomenon in Voltage Tolerant ESD Protection Circuits," *Proceedings of the 41st Annual Reliability Physics Symposium*, 2003, pp. 85–91.

29. V. Gupta, A. Amerasekera, S. Ramaswamy, and A. Tsao, "ESD-related Process Efffects in Mixed-voltage Sub-0.5 μm Technologies," *EOS/ESD Symposium Proceedings*, 1998, pp. 161–169.

30. M.-D. Ker, and W.-J. Chang, "ESD Protection Design with On-Chip ESD Bus and High-Voltage-Tolerant ESD Clamp Circuit for Mixed-Voltage I/O Buffers," *IEEE Transactions on Electron Devices*, 55(6), 2008, 1409–1416.

13 ESD Design and Optimization in Advanced CMOS SOI Technology

You Li

CONTENTS

13.1 OVERVIEW OF ESD PROTECTION IN SOI TECHNOLOGY

Silicon-on-insulator (SOI) technology builds semiconductor devices in a thin layer of silicon film on top of the dielectric-isolated silicon substrate. Comparing to the bulk silicon technology, SOI technologies demonstrate superior features including lower capacitance and leakage, elimination of latch-up, simpler manufacturing processes, and less susceptibility to soft errors. As advanced CMOS technologies have quickly entered the FinFET era, the new-generation fin-on-oxide technology offers additional benefits, thanks to the three-dimensional fin architecture and the reduced doping profile in the channel, which result in uniform conduction from fin top to bottom for better performance and expansion of transistor V_{max}/V_{min} window to enable lower operation power and better device control in manufacturing.

The advantages of SOI transistor devices with low capacitance and high drive current have accelerated the implementation of high-performance, high-frequency integrated circuits (ICs) design in SOI technologies. Particularly, the superior radio

269

frequency (RF) performances are continued to be recorded in advanced 65, 45, and 32 nm SOI CMOS technology nodes [1–3]. However, the design of effective ESD protections to the high-frequency circuits presents more unique challenges in the SOI technologies. The typical high-frequency application circuits, such as RF low-noise amplifies (LNAs) and high-speed serial (HSS) links placed at the front-end blocks of analog and digital circuits, are exposed to the danger of failure caused by ESD events. Well-designed ESD protection devices and networks are indispensable part in such circuits to guarantee product reliability and functionality.

The typical ESD protection elements available in SOI CMOS technologies include the lateral diodes, gate-non-silicide (GNS)-blocked NMOSFETs, and silicon-controlled rectifiers (SCRs). The GNS ESD NFETs are usually less attractive in advanced SOI technology nodes because of their insufficient ESD performance and severe penalty on consumed silicon area due to the requirement of large drain-side silicide blocks. Compared to the counterpart designs in bulk CMOS, ESD devices built in SOI technologies do enjoy the same benefit of low-capacitance as the SOI transistors. However, due to the presence of thin silicon film and buried oxide isolation, similar-sized ESD elements in SOI technologies usually have lower failure current and higher on-resistance because of the enhanced thermal effect and excessive self-heating. In addition, as technologies scale down, the high-performance SOI transistors having thinner gate-oxide thickness and shorter channel length are more susceptible to the ESD stress. The shrinkage of ESD design window with reduced oxide breakdown and snapback trigger voltage makes the design of robust ESD protection solutions more challenging in SOI technologies.

In this chapter, we discuss two types of primary ESD protection devices used in SOI technology: the ESD diode and SCR. The ESD performances dependency on design parameters, junction formations, and processes are investigated in detail. Optimization methodologies are also proposed based on these studies.

13.2   ESD DIODE DESIGN IN SOI TECHNOLOGY

13.2.1   OVERVIEW OF SOI DIODES

The junction diodes are frequently used in on-chip ESD protection applications because of their relatively simple structure and good performance. This is particularly true for ESD protection of low-voltage and high-frequency ICs where a low trigger voltage and low capacitance loading are required for the design of ESD protection devices. The gate-bounded and SBLK-bounded diodes are two primary ESD protection diode elements used in SOI CMOS technologies. The gate-bounded SOI diodes are well studied and utilized in various SOI technology nodes such as 90 nm [4–5], 65 nm [6], 45 nm [7,8], and 32 nm [9]. The SBLK-bounded diode is an alternative SOI diode design and formed by using the silicide blocking (SBLK) layer instead of polysilicon gate between diode anode and cathode region. It was first introduced in [10] for the advantages of lower leakage and lower capacitance. Figure 13.1 shows the cross sections of typical P+/N-body (PNB) gate-bounded and SBLK-bounded ESD diodes in SOI technology. Both diodes are built in a thin silicon

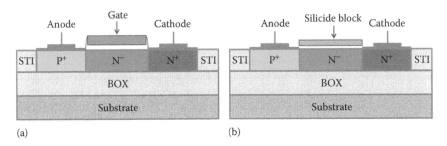

FIGURE 13.1   Cross sections of typical (a) gate-bounded and (b) SBLK-bounded ESD diode in SOI technology.

film on top of the buried oxide (BOX) and have the anode and cathode implantation touching the surface of BOX layer. The P+/N-body junctions are formed in a lateral direction only, and there are no vertical junctions underneath the anode region of both diodes. The anode-to-cathode spacing (SAC) is defined by the length of poly-silicon gate in gate-bounded and silicide-block region in SBLK-bounded SOI diode, respectively. Note that for the gate-bounded diode, the gate terminal can be either floating or tied to cathode. The configuration of connecting gate to anode is not desirable because of the high gate-to-body capacitance. The gate floating is a pre-ferred design because of the benefits in capacitance reduction and enhanced oxide robustness under CDM event [10].

In a typical diode-based ESD protection scheme, the double-diode structure having one diode connecting from the I/O pad to power rail and the other from ground rail to I/O pad is implemented with power clamps between the power and ground rails to provide whole-chip current discharge paths. For a robust ESD protection design, the diodes must sustain high ESD current and clamp the voltage to a safe region, such as below the drain breakdown or gate-oxide breakdown voltage of the protected transistors. Two key ESD metrics, failure current and on-resistance, are typically used to evaluate the performance of ESD diodes. On the other hand, during the normal circuit operation, it is desirable to have ESD diodes with low para-sitic capacitance and low leakage current. Due to the presence of buried oxide in SOI technology, the ESD SOI diodes do enjoy the benefit of low-capacitance but endure a penalty of excessive self-heating since the insulating property of BOX layer reduces the heat dissipation capability. The trade-off between superior ESD performance and minimized capacitance loading still remains a main challenge for design of robust ESD diodes in SOI technologies.

13.2.2 ESD Performance of SOI Diodes

The ESD performance of SOI diodes are characterized with standard transmission line pulse (TLP) testing. The devices are stressed by the pulses generated from the TLP system with a rise time of 10 ns and pulse width of 100 ns. A DC leakage test is performed after each TLP pulse, with a 1 μA forward current injection applied at the anode of diode, and the cathode is grounded. A 10% shift in DC voltage is defined

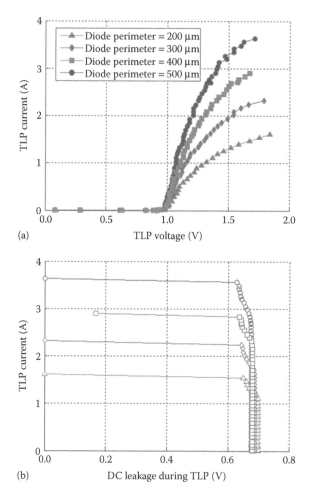

FIGURE 13.2 (a) 100 ns TLP I–V as well as (b) DC leakage results of PNB gate-bounded SOI diode with perimeter variation.

as the ESD failure point of diode. Figure 13.2 shows the 100 ns TLP I–V as well as DC leakage results of PNB gate-bounded SOI diode with perimeter scaling from 200, 300, 400, to 500 μm. All diodes have the same design of finger width and metal wiring scheme. The perimeters are varied by increasing the number of diode fingers. The DC leakage results show that all diodes start with a soft leakage shift followed by a hard failure.

The failure current and on-resistance of PNB gate-bounded SOI diode versus perimeter variation are extracted and plotted in Figure 13.3. It is observed that the on-resistance is inversely proportional and failure current is directly proportional to the diode perimeter. Both of them scale reasonably with the device perimeter variation. A failure current of ~7.2 mA/μm per perimeter is achieved for the gate-bounded SOI diode. The on-resistance reduces from 0.4 to 0.19 Ω with diode perimeter increasing

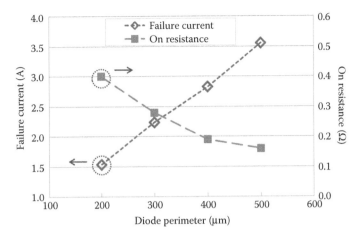

FIGURE 13.3 Extracted failure current and on-resistance of PNB gate-bounded SOI diode versus diode perimeter.

from 200 to 400 μm. The data also reveals that for diode having the largest perimeter of 500 μm, the on-resistance drops less. This is because at larger device perimeter, the wiring resistance becomes a significant part and starts to dominate the overall resistance of diode.

To achieve an optimal diode design, several key device dimensions are studied to understand their effects to the ESD performance of SOI diode. The failure current and on-resistance of PNB gate-bounded diode against various anode lengths and anode-to-cathode spacing are plotted in Figure 13.4. It can be seen that both failure current and on-resistance are not sensitive to the length of anode. This is because the PNB SOI diode has anode implantation touching the surface of BOX layer and the P+/N-body junction is formed laterally only. Unlike the counterpart device in bulk technology, increasing the length of anode region does not impact the total junction area in SOI diode and thus the ESD performance. However, smaller anode length will result in better failure current per silicon area. The effect of anode-to-cathode spacing in SOI diode is consistent with the trend of bulk diode design, increasing the SAC results in a slightly decreasing of failure current and significant on-resistance increase of the gate-bounded SOI diode. The smaller SAC design in SOI diode is preferred to achieve adequately lower clamping voltage.

The measured total capacitance of PNB gate-bounded SOI diode with perimeter variation is shown in Figure 13.5 at different reverse-bias voltages. The total capacitance includes both the front-end-of-line (FEOL, i.e., silicon and below) and back-end-of-line (BEOL, i.e., contact and above) capacitance. It is noted that the capacitance of diode is bias dependent, and a larger reverse voltage results in a lower total diode capacitance. Good capacitance scaling is observed for the gate-bounded SOI diode with a capacitance of 0.45 fF/μm per perimeter at the 0 V bias. The measurement data is also evident that the capacitance reduction increases as the reverse bias increases. This is possibly due to the gate biasing effect. For gate-floating configuration, when the anode is biased at −3 V and cathode is grounded, the leakage

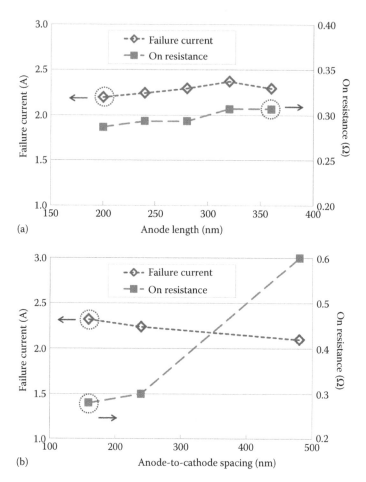

FIGURE 13.4 Extracted failure current and on-resistance of PNB gate-bounded SOI diode versus (a) anode length and (b) anode-to-cathode spacing.

current through the gate makes the gate slightly negatively biased, thus further reducing the total diode capacitance. Due to the removal of gate to anode/cathode overlap capacitance, the SBLK-bounded SOI diode consistently achieves ~15%–20% capacitance reduction than the gate-bounded design.

13.2.3 ESD PERFORMANCE OF FIN-BASED SOI DIODES

As SOI technology progresses from planar devices to vertical structures such as the three-dimensional fin-based MOSFET (FinFET) transistors [11,12], the tradition designs of ESD devices in planar region are no longer compatible because of the implementation of fin-patterning steps in FinFET fabrication processes. To build an ESD diode in FinFET technology, either a group of new designs need to be explored

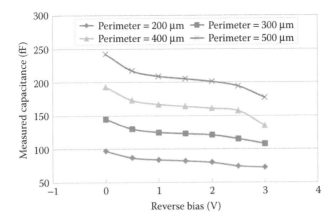

FIGURE 13.5 Measured total capacitance of gate-bounded SOI diode under different reverse bias voltages.

by using the patterned fins or a planar region need to be reserved to form the same kinds of diode design used in past technology nodes. Figure 13.6a shows the illustration of fin-based ESD gate-bounded diode in an SOI FinFET technology. The cross sections of diode cut view along and against the fins are also shown in Figure 13.6b and c. The fin-based ESD diode is a straightforward variation of FinFET transistor with P⁺ and N⁺ regions defined on the same fin to form the anode and cathode of diode. It employs the same fabrication steps as FinFET transistors, and they share with most of the semiconductor components such as the gate stack, epitaxial growth, junctions, and silicide contacts. To ensure a low wiring resistance, the standard full BEOL metal levels need to be processed. As fin-based diode builds the structure based on a patterned fin array, the critical design dimensions affecting the ESD

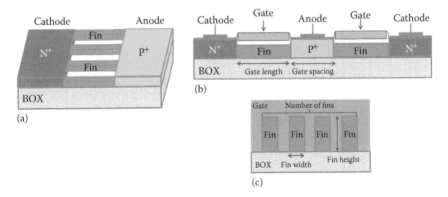

FIGURE 13.6 Illustration of fin-based ESD gate-bounded diode in an SOI FinFET technology (all gates are omitted) (a) and cut views of (b) along and (c) against the fins.

performance are the fin width, fin height, number of fins, gate length, and gate spacing, and they are also labeled in the Figure 13.6b and c to facilitate discussion.

Figure 13.7 shows the 100 ns TLP I–V as well as DC leakage results of fin-based ESD gate-bounded SOI diode with perimeter scaling from 72 to 144 μm. The diode perimeter is increased by increasing the total number of fins. To have a fair comparison with diode designed in the planar region, the perimeter of fin-based diode is calculated as Perimeter = 2 × (Number of fin) × (Fin width). The DC leakage data shows that both diodes have the soft failure with a gradual leakage current shifting. Good scaling of failure current and on-resistance with diode perimeter are observed. An excellent normalized failure current of ~13.5 mA/μm per perimeter is recorded for the fin-based diode design. The achievement of outstanding failure current is particularly due to the unique device features available in the FinFET architecture, such as the large spacing between each fin (therefore, the large device area), the three-dimensional gate wrapped around the fin, and the

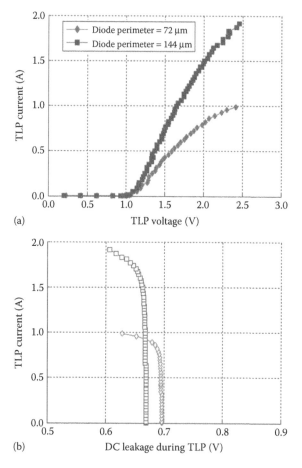

(a)

(b)

FIGURE 13.7 (a) 100 ns TLP I–V as well as (b) DC leakage results of fin-based gate-bounded SOI diode with perimeter variation.

extensive BEOL wiring used to connect all fins, all of which acting as effective heat sinks to greatly help dissipate the heat and improve the failure current of fin-based diode. However, the excessive gates and wiring capacitance are introduced for diode with this design.

13.2.4   FIN-BASED SOI DIODES OPTIMIZATION

The silicon film thickness of the starting SOI substrate (Tsi) has been varied to investigate its effect to the ESD performance of fin-based diode. Figure 13.8 shows the 100 ns TLP results comparison of fin-based diodes built on the standard silicon film thickness (1x of Tsi) and 1.4 times thicker thickness (1.4x of Tsi) with different perimeters. It is shown that the performance benefits of fin-based diodes on thicker silicon film are not significant. Only ~10% TLP failure current improvement is observed for the diodes having 40% thicker silicon. This is because the increase of silicon film thickness only benefits the fin regions underneath the gate of diode, which is a small portion of the entire device. On the other hand, to form proper silicide contacts, the anode and cathode regions of fin-based diode always have the selectively grown silicon. They are not affected by the thickness change of silicon film and thus do not contribute to any failure current improvements. There is no observed on-resistance reduction for the fin-based diodes with thicker silicon film. This is because the patterned fin arrays need to rely on the rules of fin pitch and gate spacing, which results in a much larger consumption of silicon area. The wiring resistance increases considerably and starts to dominate the total resistance of fin-based diode. The ESD performance boost from fin thickness change is further masked by the increased wiring resistance.

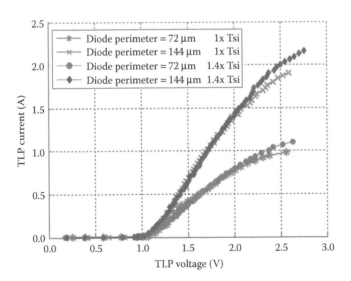

FIGURE 13.8   100 ns TLP I–V data of fin-based gate-bounded SOI diodes with silicon film thickness variation.

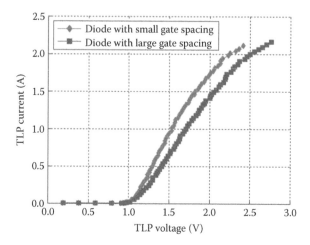

FIGURE 13.9 100 ns TLP I–V data of fin-based gate-bounded SOI diodes with gate spacing variation.

To reduce the total device area, the effects of gate spacing to the ESD performance of fin-based diode are investigated. Figure 13.9 shows the 100 ns TLP results comparison of fin-based diodes using different gate space. They all have the same perimeter and silicon film thickness. A ~20% of on-resistance reduction is observed for diode with the smaller gate spacing. There is no significant decrease in failure current due to the reduced gate space design. For both diodes, the failure currents are in a similar level of 14.5 mA/µm per perimeter. This is indicated that the reduced gate spacing in fin-based diode design does not result in a penalty of heat dissipation and ESD performance degradation; therefore, it is the most optimized design because of the reduction of total device area.

A clear disadvantage of the fin-based diode is the significant loss of silicon volume because the design is limited by the rules of fin pitch and gate spacing. To compare the ESD performance, the planar-based ESD diodes are fabricated onto the same FinFET SOI technology by reserving a planar silicon region. Extra process steps and masks are necessary to achieve the planar region on the same wafer. Figure 13.10 shows the normalized TLP results per diode perimeter and per diode area to compare the fin-based and planar-based design approach. From this comparison, one may find the fin-based diode appealingly shows a much higher failure current per perimeter (e.g., 14.5 mA/µm vs. 6.2 mA/µm). However, when the TLP results are normalized by silicon area to reflect the on-wafer cost of ESD devices, the performance of planar-based diode jumps out. It shows a failure current of 16 mA/µm² per area, whereas the fin-base diode shows only 2.4 mA/µm². This is not surprising given the large fin spacing required for the fin-patterning steps. The area advantage of the planar-based diode is clearly demonstrated with a failure current benefit of almost seven times higher than the fin-based design.

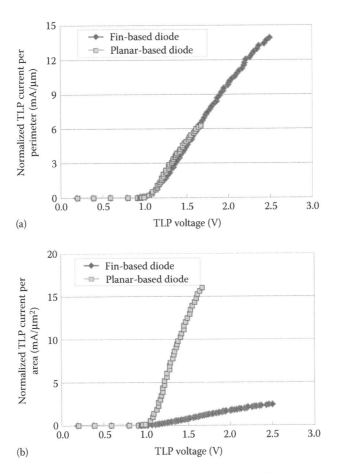

FIGURE 13.10 100 ns TLP I–V data of fin-based versus planar-based gate-bounded SOI diodes normalized by (a) perimeter and (b) silicon area.

13.3 ESD SCR DESIGN IN SOI TECHNOLOGY

13.3.1 SCR ON SOI VERSUS BULK TECHNOLOGY

SCR is another widely used ESD structure in advanced CMOS technologies for on-chip ESD protection. Advantages of SCR devices include high-current conduction, low DC leakage, and low parasitic capacitance. In addition, when used in I/O ESD protection, unlike the diode-based scheme, SCRs can discharge ESD current from I/O to ground rail directly, without relying on the low-power bus resistance to achieve adequate clamping voltage. Extensive silicon results of SCR design and optimization in advanced bulk CMOS technologies have been presented in various publications [13–15]. However, due to the unique device features in SOI, the same design of SCR implemented in bulk CMOS cannot be transferred to SOI technology directly [16–19]. The SCR device built in an SOI technology has the unique characteristics including:

1. The lateral PNPN structure formed in a thin silicon film
2. No existing parasitic devices to substrate
3. Lateral body contact scheme for cross-coupled PNP and NPN
4. Isolation between external triggering circuits and main SCR

Figure 13.11 shows the cross sections of an SOI and a corresponding bulk SCR device for comparison. Because of thin silicon film thickness, the N+ and P+ diffusion in SOI SCR device can penetrate through the well regions and reach the surface of BOX. This results in a butting anode and cathode junction to build the lateral PNPN structure. In addition, due to the presence of buried oxide layer, there are no parasitic elements to substrate formed in the SOI SCR device. In comparison, a vertical PNP bipolar formed by the P+ anode, N-well, and P-substrate exists in the bulk SCR structure. The SCR design parameter SAC (spacing between anode and cathode) is defined as the length of silicide-blocked (SBLK) region between P+ anode and N+ cathode in SOI SCR and the length of shallow trench isolation (STI) region in bulk SCR device, respectively. The base widths of the cross-coupled SCR PNP and NPN transistors are determined mainly by a function of SAC parameter. In bulk SCR, as the bipolar current must pass underneath the STI region, the thickness of STI oxide need also be considered. For SOI SCR, the SAC (e.g., length of SBLK region) defines the sum of PNP and NPN bipolar base widths straightly. However, as the anode and cathode implantation in SOI SCR touches the BOX surface, there are no N-body and P-body well resistors (e.g., base resistance of the PNP and NPN bipolar) formed

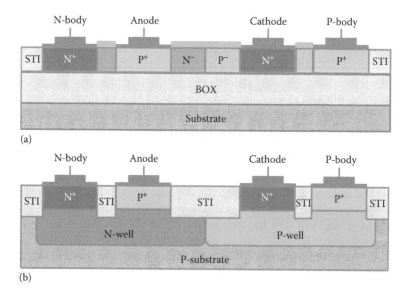

(a)

(b)

FIGURE 13.11   Cross sections of (a) an SOI SCR and (b) a bulk SCR device.

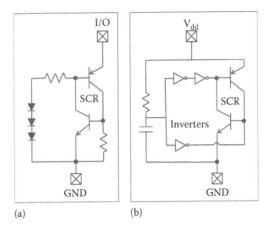

FIGURE 13.12 Typical circuit schematics of (a) a DTSCR and (b) an RCSCR device.

underneath the anode and cathode junction. In contrast, the body resistances in SOI SCR device need to be formed in the third dimension.

In general, the triggering of native SCR devices are breakdown dominated, and the trigger voltages are too high for circuits designed in advanced CMOS technologies. Therefore, the external triggering techniques are usually integrated into main SCR to reduce the trigger voltage down to an acceptable level. The typical triggering circuits include a diode string and an RC network. Figure 13.12 shows the circuit schematics of a diode-triggered SCR (DTSCR) and an RC-triggered SCR (RCSCR) in SOI technology. For DTSCR device, the external trigger diodes are connected to the N-well base of SCR PNP transistor. This type of DTSCR design is typically used in low-capacitance and low-voltage applications, such as the I/O ESD protection for RF LNAs and HSS links circuits. The RCSCR device consists of an RC network and the inverter chains to form a dual-path control circuit. The RC network is responsible for detecting the ESD events and initiating SCR triggering process. The inverter chains are connected to the N-well base of PNP and P-well base of NPN bipolar, respectively, to supply triggering current for SCR under ESD conditions. During normal operation, the inverter chains hold the N-well at the same bias of SCR anode and pull the P-well down to ground with SCR cathode. There are no forward-biased junctions in the SCR and thus the leakage is minimized. The RCSCR devices are preferred ESD solutions in leakage-sensitive applications such as power supply clamp for the battery-powered mobile chips. Note that to avoid any latching issues, the holding voltage of RCSCR used in power pin protection must be greater than the VDD supply voltage by a safe margin.

13.3.2 ESD Performance of SOI SCR

To characterize the ESD performance of SCR design in SOI technology, the SOI DTSCR devices are stressed by the pulses generated from the TLP system with a rise time of 10 ns and pulse width 100 ns. The DC leakage testing is performed after

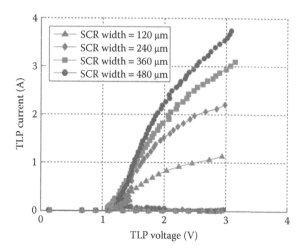

FIGURE 13.13   100 ns TLP I–V data of SOI DTSCR devices with width variation.

each TLP pulse with a 1.0 V bias voltage across the anode and cathode of SCR. The failure current is defined by the last data point in the TLP I–V curve before the DC leakage current shifts five times or more. The holding voltage is determined by extrapolating the linear portion of the TLP I–V data to the voltage point where the TLP current is zero. The failure current and holding voltage are two key ESD performance metrics for evaluation of SCR design. Due to the latch-up nature of SCR device, once successfully triggered, the SCR maintains the on-state by itself, thus those ESD performance metrics are independent to the different external triggering techniques implemented in SCR.

Figure 13.13 shows the 100 ns TLP I–V data of SOI DTSCRs with device width variation from 120 to 480 μm. The SCR width varies by increasing the number of total SCR fingers. Two trigger diodes are used in the external triggering circuit. As shown in Figure 13.13, the failure current of SOI DTSCR device scales well with SCR width and a normalized failure current of ~8.2 mA/μm per width are achieved. Although these DTSCR devices all have a large number of parallel fingers (e.g., more than 20 fingers), there are no multi-finger turn-on issues observed. Once triggered on, all the parallel fingers snapback to a holding voltage of ~1.2 V. Hard failures are observed for all SCR devices.

Key factors affecting SCR triggering and ESD performance are the SCR bipolar gains, SCR bipolar base resistances, and effective triggering current. In SOI SCR design, the bipolar characteristics are strongly impacted by several critical design dimensions such as the anode-to-cathode spacing and body-contact spacing. Those design factors are studied with 100 ns TLP testing results to understand their effects to the triggering behavior and ESD performance of SCR in SOI technology.

The gains of SCR cross-coupled PNP and NPN transistors, namely the bipolar betas, are mainly determined by their base widths. In SOI SCR, the SCR anode-to-cathode spacing defines the sum of PNP and NPN base widths. Figure 13.14 shows the extracted failure current and holding voltage of SOI DTSCRs with SAC varying

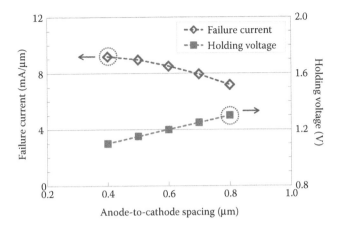

FIGURE 13.14 Extracted failure current and holding voltage of SOI DTSCR devices with SAC variation.

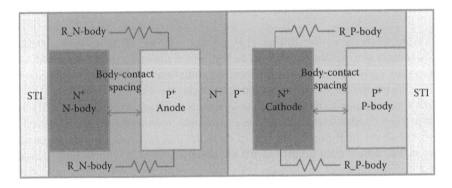

FIGURE 13.15 Illustration of top view of the SOI SCR device.

from 0.4 to 0.8 μm. All SCRs have a total device width of 360 μm. The DTSCR is an actively triggered SCR device. As shown in Figure 13.14, with smaller anode-to-cathode spacing, the holding voltage scales down and the failure current scales up, due to the higher bipolar gains. At the smallest SAC, an excellent failure current of ~9.2 mA/μm per width and a holding voltage of ~1.1 V can be achieved for the SOI DTSCR device. However, this results in a penalty of DC leakage current increase due to the reduced bipolar base width.

The body-contact spacing is another key design dimension for SCR built in SOI technology. Since both the N$^+$ and the P$^+$ implantation touch the buried oxide, there are no N-body and P-body resistors formed underneath the SCR anode and cathode regions. Instead, the base resistances of the cross-coupled PNP and NPN transistors need to be built in the third dimension, as shown in Figure 13.15. The body-contact spacing is defined as the space of N-body contact to SCR anode and P-body contact to SCR cathode terminals. The larger body-contact spacing results in the increased bipolar base resistance.

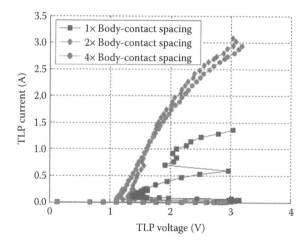

FIGURE 13.16   100 ns TLP I–V data of SOI DTSCR devices with body-contact spacing variation.

Figure 13.16 shows the 100 ns TLP I–V data of SOI DTSCRs with different body-contact spacing design, all with an SAC of 0.6 μm and a total device width of 360 μm by consisting of a large number of parallel fingers. The base resistance of SCR NPN bipolar (R_P-body) is varied by changing the spacing between the P-body contact and cathode terminal of SCR. As shown in Figure 13.16, all the SCRs are able to trigger and sustain current. However, visible multi-finger triggering behavior (e.g., not all the SCR fingers are turned on at the first snapback), higher triggering current, and much lower failure current are observed for SCR with 1x body-contact spacing. This indicates inadequate body resistance to sustain the snapback of all SCR fingers. By increasing the body-contact spacing to 2x, excellent SCR triggering behavior with a reduced trigger current of ~50 mA and smooth I–V curve after the trigger point is shown. Further increasing the body-contact spacing to 4x no longer improves the SCR triggering and ESD performance, except a slightly deeper snapback is seen, suggesting a lower holding voltage.

13.3.3   JUNCTION ENGINEERING

In an early SCR development phase, the characteristics of SCR cross-coupled PNP and NPN transistors are sensitive to the anode/cathode junctions formed in the SCR. However, little work has been done to investigate the relationship between triggering/performance of SCR devices and the junction engineering, especially in the advanced SOI technologies. Junction formation can result in distinct bipolar characteristics and thus affects the SCR triggering behavior and its ESD performance.

Figure 13.17 shows the 100 ns TLP I–V as well as DC leakage results of SOI DTSCR with process variations in terms of SCR cathode N$^+$ implant dosage.

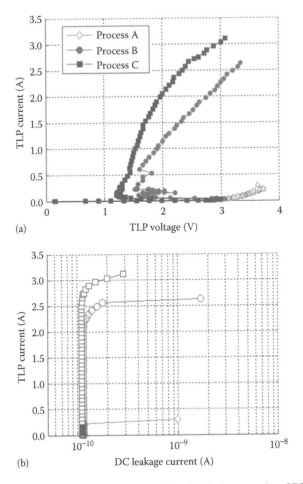

FIGURE 13.17   (a and b) 100 ns TLP I–V as well as DC leakage results of DTSCR devices with cathode N⁺ implant dosage variation.

The implantation energy is high enough to form the butting cathode junctions in all the processes. All SCRs have an SAC of 0.6 µm, a total width of 360 µm and two external trigger diodes. The N⁺ implant dosage at SCR cathode region increases from process A to C. As shown in Figure 13.17, the SCR with process A with the lowest N⁺ implant dosage fails to trigger. By increasing the N⁺ implant dosage, the SCR with process B is able to trigger and sustain current. However, the zig-zag curve after the triggering point indicates the multi-finger triggering issues, and the SCR has a lower failure current and higher on-resistance. This can be attributed to the NPN bipolar having a lower current gain [20]. The SCR with process C, with an even higher N⁺ implant dosage, has a smooth I–V curve after the triggering point and an excellent failure current of ~8.3 mA/µm per width is observed.

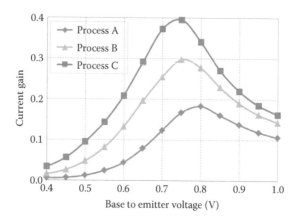

FIGURE 13.18   Measured NPN current gains of SOI SCR with cathode N⁺ implant dosage variation.

Both standard bipolar gain and customer bipolar gain data under DC condition are measured with a four-terminal SCR structure. The four-terminal SCR is designed with all four terminals of device wired to the pads for wafer-level testing. Figure 13.18 shows the representative plot of current gain data of the NPN bipolar (e.g., N-body, P-body, and cathode terminals of the four-terminal SCR structure) with different SCR cathode N⁺ implant dosage A, B, and C. The voltage bias is applied to the base (SCR P-body) and emitter (SCR cathode) terminals, and the collector (SCR N-body) is grounded. As shown in Figure 13.18, the NPN with implant dosage C has the highest current gain under all the bias condition. For all bipolar, the current gain peak when the base-to-emitter voltage is between 0.7 and 0.8 V. The strong sensitivity of NPN bipolar gain to SCR cathode N⁺ implant dosage is caused by two factors. First, a higher N⁺ dosage increases the doping of NPN emitter (SCR cathode) near the base–emitter junction and thus improving the NPN emitter efficiency. Second, the N⁺ cathode diffuses further into the P-body region as dosage increases, moving the NPN base–emitter junction closer to the N-body/P-body junction and thus reducing the physical base width of NPN bipolar.

The triggering behavior and ESD performance of RCSCR devices in SOI technology with the same process variations are plotted in Figure 13.19. The RCSCRs have the same SAC of 0.6 μm and total width of 360 μm as DTSCR devices. It is shown that the triggering of RCSCRs is less sensitive to the process variation compared to the DTSCR devices. Even the SCR fabricated in process A with the lowest N⁺ implant dosage at cathode and, therefore, the lowest NPN bipolar gain has no problem to trigger and sustain current. This is because both PNP and NPN transistors in the RCSCR device are actively biased by the external RC triggering network, and the bipolar gain and bipolar base resistance are no longer the gating factors for the SCR triggering. It is concluded that the RCSCR device is less sensitive to the formation of junction compared to the DTSCR device because of their different triggering

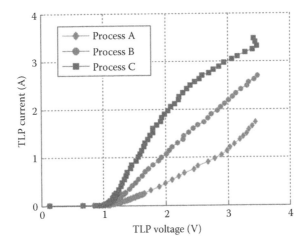

FIGURE 13.19   100 ns TLP I–V data of SOI RCSCR devices with cathode N⁺ implant dosage variation.

techniques. However, the RCSCR with lower N⁺ implant at cathode dosage shows a lower failure current and higher on-resistance.

13.3.4   Halo/Extension Effects

Extension and halo implants are employed in regular MOSFET fabrication processes for transistor performance improvement. The extension implant utilizes the same type of doping as the transistor source/drain implant. It performs the shallow and heavily doped implantation to minimize device resistance. The halo implant is tilted implantation and counter doped to the source/drain implant of transistor to minimize the short channel effect. To understand their effects to ESD devices, including the triggering and ESD performance of SCR, the SOI DTSCR devices with an additional extension/halo implants are studied.

In the design experiments, the SCR cathodes are formed by a two-step implantation. The normal SCR cathode implant is implemented at the first step to build a butting cathode junction touching the BOX surface and followed by an additional NFET extension and/or halo implants. Figure 13.20 shows the 100 ns TLP I–V and DC leakage results of SOI DTSCR devices with SCR using the two-step cathode implantation. The TLP data of DTSCR having normal SCR cathode implant is also plotted for the comparison. As shown in Figure 13.20, the extension implant does not help the smooth triggering of DTSCR, and a slightly lower failure current and higher on-resistance is observed.

The peak NPN gains of SCR with the additional NFET extension and halo implants are also measured and extracted in Figure 13.21 at various collector-to-base (e.g., N-body and P-body terminals of the four-terminal SCR structure) bias conditions. For all the designs, the peak current gain increases with larger collector-to-base bias voltage. The SCRs with NFET extension implant have the

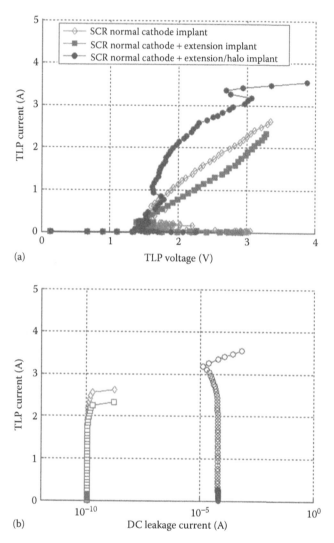

FIGURE 13.20 (a) 100 ns TLP I–V as well as (b) DC leakage results of SOI DTSCRs with extension and extension/halo implants.

highest NPN bipolar gains. Compared to SCR having the normal cathode implant, the peak NPN current gain increases from 0.3 to 0.46 at the collector-to-base bias voltage of 2 V. This is due to the additional highly doped N⁺ implant at SCR cathode, which improves emitter efficiency and reduces physical base width of the NPN bipolar. However, as shown in Figure 13.20, the smooth triggering of SCR is not improved by the extension implant, and a lower failure current and higher on-resistance is observed. It is believed that only the surface portion of the SCR turns

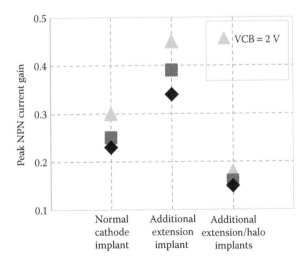

FIGURE 13.21 Peak current gains of NPN in SCR with extension and extension/halo implants at various collector-to-base bias conditions.

on, because the added shallow and highly doped N-type implantation results in a non-uniform NPN bipolar gain in the vertical direction of SOI film and the surface region of NPN have higher current gain due to its higher dosage and narrower base width.

On the other hand, by implementing both NFET extension and halo implants, the peak NPN current gains drop significantly due to the P$^+$ counter doping (e.g., the peak bipolar gain of SCR with extension/halo implants is only 0.18 at a collector-to-base bias of 2 V). The trigger voltage of SCR in this design reduces to ~2.2 V, and the DC leakage current increases significantly to the range of several μAs, as shown in Figure 13.20. The increased leakage current is caused by the affected SCR PNP bipolar due to the halo effect. This is confirmed by the current components measured at each terminal of the PNP bipolar (e.g., anode, N-body, and P-body of the four-terminal SCR structure) under DC condition. As shown in Figure 13.22, a high-leakage current path is formed between the emitter (SCR anode) and collector (SCR P-body) terminals of PNP bipolar even under a small emitter-to-base bias voltage. The current gain of PNP is defined as I_pbody/I_nbody. The leakage path results in a lager PNP bipolar gain and higher DC leakage current for SCR with the additional extension/halo implants.

The capacitance of DTSCR with various junction engineering designs is measured and plotted in Figure 13.23. It is noted that the capacitance of DTSCR is not sensitive to the junction formation. This is due to the fact that SCR anode/cathode junction capacitors are connected in series with the N-body/P-body capacitor, which has a lower value and plays a dominant role in total capacitance of SCR. For all the designs, the capacitance scales well with the SCR width, and a capacitance of 0.31 fF/μm is achieved.

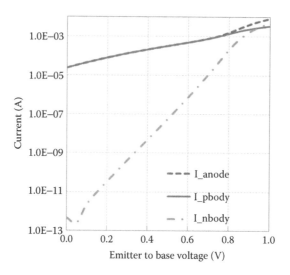

FIGURE 13.22 Measured current components of PNP bipolar of SOI DTSCR with extension/halo implant.

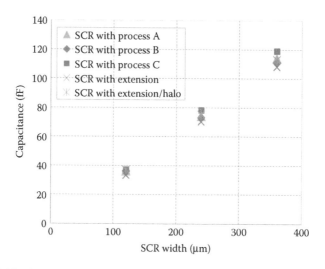

FIGURE 13.23 Measured capacitance of SOI DTSCR with various junction engineering designs.

13.4 SUMMARY

SOI technology has proven to be attractive for high-performance, high-frequency circuits design because of its performance advantages over similar bulk CMOS technologies. However, due to the use of thin silicon film and the presence of buried-oxide isolation, similar-sized ESD elements employed in SOI technologies usually have lower ESD performance. In addition, with aggressive technology scaling and drive for high performance, the shrink of ESD design window makes the design of robust ESD protection devices in SOI CMOS more challenging. The ESD performance versus capacitance trade-off should be carefully considered when investigating ESD protection solutions in SOI technologies.

Design and optimization of two primary ESD elements, the ESD SOI diode and SCR, in advanced nanometer SOI CMOS technologies are studied and presented in this chapter. In the typical diode-based ESD protection scheme, a double-diode structure is implemented with the power clamps between power and ground rails to provide whole-chip ESD current discharge. The ESD performance of PNB gate-bounded and SBLK-bounded SOI diodes in planar region is first investigated with diode anode length and anode-to-cathode spacing variation. A failure current of ~7.2 mA/μm and capacitance of 0.45 fF/μm per perimeter is achieved for the gate-bounded SOI diode. Due to the removing of gate to anode/cathode overlap capacitance, the SBLK-bounded SOI diode consistently achieves ~15%–20% capacitance reduction. As advanced CMOS technologies have quickly entered the FinFET era, the ESD performance of fin-based diode in SOI FinFET technology is also evaluated. Key device variations, including the silicon film thickness and the gate spacing, are investigated for performance optimization. The failure current of ESD diodes designed in fin-based and planar-based approaches is compared by the normalized TLP results per perimeter and per silicon area, respectively. The fin-based diode shows a higher failure current per perimeter than planar diode. However, when both diodes are normalized by silicon area to represent the on-wafer cost of ESD devices, the fin-based diode gives much lower failure current per area due to the large fin spacing required for the fin patterning steps. The area advantage of the planar-based diode is clearly demonstrated with a failure current benefit of almost seven times higher than the fin-based design.

SCR is another type of ESD device widely used in advanced CMOS technologies for on-chip ESD protection. Advantages of SCR devices include high-current conduction, low DC leakage, and low parasitic capacitance. In addition, when used in I/O ESD protection, unlike the diode-based scheme, SCRs can discharge ESD current from I/O to GND rail directly, without relying on the low-power bus resistance to achieve adequate clamping voltage. However, due to the presence of buried oxide, the same SCR design implemented in bulk CMOS cannot be transferred to SOI directly. An SCR built in an SOI has the unique characteristics compared to the counterpart device in bulk CMOS. The ESD performance of SOI DTSCR and RCSCR is first studied with anode-to-cathode spacing and body-contact spacing variation. A failure current of ~8.2 mA/μm and capacitance of 0.31 fF/μm per width is achieved for the SOI DTSCR device. The effect of junction engineering in terms of different SCR cathode N^+ implant dosage and the impact of NFET extension and halo implants on SCR triggering behavior and ESD performance are also investigated. The RCSCR device is less sensitive to the formation of junction compared to the DTSCR because of their different triggering techniques. The extension implant has surprising negative impact on the failure current of SCR devices due to the non-uniform bipolar gain causing partial turn-on of the SCR. The halo implant results in a penalty of significant leakage current increase. Higher implant dosage improves the SCR performance with smooth triggering and increased failure current due to the higher SCR bipolar gains.

REFERENCES

1. S. Lee, J. Kim, D. Kim, B. Jagannathan, C. Cho, J. Johnson, B. Dufrene, N. Zamdmer, L. Wagner, R. Williams, D. Fried, K. Rim, J. Pekarik, S. Springer, J. Plouchart, and G. Freeman, "SOI CMOS Technology with 360 GHz fT NFET, 260 GHz fT PFET, and Record Circuit Performance for Millimeter-Wave Digital and Analog System-on-Chip Applications," in *Proceedings of the IEEE Symposium on VLSI Technology*, pp. 54–55, 2007.

2. S. Lee, B. Jagannathan, S. Narasimha, A. Chou, N. Zamdmer, J. Johnson, R. Williams, L. Wagner, J. Kim, J.-O. Plouchart, J. Pekarik, S. Springer, and G. Freeman, "Record RF Performance of 45-nm SOI CMOS Technology," in *Proceedings of the IEEE International Electron Devices Meeting (IEDM)*, pp. 255–258, 2007.

3. S. Lee et al., "Advanced Modeling and Optimization of High Performance 32 nm HKMG SOI CMOS for RF/Analog SoC Applications," in *Proceedings of the IEEE Symposium on VLSI Technology*, pp. 135–136, 2012.

4. A. Salman, S. Beebe, M. Pelella, and G. Gilfeather, "SOI Lateral Diode Optimization for ESD Protection in 130 nm and 90 nm Technologies," in *Proceedings of the IEEE International EOS/ESD Symposium*, pp. 421–427, 2005.

5. V. Chen, S. Beebe, E. Rosenbaum, S. Mitra, C. Putnam, and R. Gauthier, "SOI Poly-Defined Diode for ESD Protection in High Speed I/Os," in *Proceedings of the IEEE International Reliability Physics Symposium (IRPS)*, pp. 635–636, 2006.

6. S. Mitra, C. Putnam, R. Gauthier, R. Halbach, C. Seguin, and A. Salman, "Evaluation of ESD Characteristics for 65 nm SOI Technology," in *Proceedings of the IEEE International SOI Conference*, pp. 21–23, 2005.

7. J. Li, S. Mitra, H. Li, M. Abou-Khalil, K. Chatty, and R. Gauthier, "Capacitance Investigation of Diode and GGNMOS for ESD Protection of High Frequency Circuits in 45 nm SOI CMOS Technologies," in *Proceedings of the IEEE International EOS/ESD Symposium*, pp. 228–234, 2008.

8. S. Cao, A. Salman, S. Beebe, M. Pelella, J. Chun, and R. Dutton. "ESD Device Design Strategy for High Speed I/O in 45 nm SOI Technology," in *Proceedings of the IEEE International EOS/ESD Symposium*, pp. 235–241, 2008.

9. S. Mitra, R. Gauthier, C. Putman, R. Halbach, and C. Seguin, "Impact of Stress Engineering on High-k Metal Gate ESD Diodes in 32 nm SOI Technology," in *Proceedings of the IEEE International EOS/ESD Symposium*, pp. 1–7, 2009.

10. C. Putnam, M. Woo, R. Gauthier, M. Muhammad, K. Chatty, C. Seguin, and R. Halbach, "An Investigation of ESD Protection Diode Options in SOI," in *Proceedings of the IEEE International SOI Conference*, pp. 24–26, 2004.

11. A. Paul et al., "Comprehensive Study of Effective Current Variability and MOSFET Parameter Correlations in 14 nm Multi-Fin SOI FINFETs," in *Proceedings of the IEEE International Electron Devices Meeting (IEDM)*, pp. 1351–1354, 2013.

12. C.H. Lin et al., "High Performance 14 nm SOI FinFET CMOS Technology with 0.0174 μm^2 embedded DRAM and 15 Levels of CU Metallization," in *Proceedings of the IEEE International Electron Devices Meeting (IEDM)*, pp. 381–383, 2014.

13. M. Mergens, C. Russ, K.G. Verhaege, J. Armer, P.C. Jozwiak, R. Mohn, B. Keppens, and C.S. Trinh, "Diode-Triggered SCR (DTSCR) for RFESD Protection of BiCMOS SiGe HBTs and CMOS Ultra-Thin Gate Oxides," in *IEEE International Electron Devices Meeting (IEDM)*, pp. 2131–2134, 2003.

14. J. Di Sarro, K. Chatty, R. Gauthier, and E. Rosenbaum, "Study of Design Factors Affecting Turn-On Time of Silicon Controlled Rectifiers (SCRs) in 90n and 65 nm Bulk CMOS Technologies," in *Proceedings of the IEEE International Reliability Physics Symposium (IRPS)*, pp. 163–168, 2006.

15. R. Gauthier, M. Abou-Khalil, K. Chatty, S. Mitra, and J. Li, "Investigation of Voltage Overshoots in Diode Triggered Silicon Controlled Rectifiers (DTSCRs) Under Very Fast Transmission Line Pulsing (VFTLP)," in *Proceedings of the IEEE International EOS/ESD Symposium*, pp. 1–10, 2009.
16. O. Marichal, G. Wybo, B. Van Camp, P. Vanysacker, and B. Keppens, "SCR Based ESD Protection in Nanometer SOI Technologies," in *Proceedings of the IEEE International EOS/ESD Symposium*, pp. 372–379, 2005.
17. C. Entringer, P. Flatresse, P. Galy, F. Azais, and P. Nouet, "Partially Depleted SOI Body-Contacted MOSFET-Triggered Silicon Controlled Rectifier for ESD Protection," in *Proceedings of the IEEE International EOS/ESD Symposium*, pp. 166–171, 2006.
18. M.P.J. Mergens, O. Marichal, S. Thijs, B. Van Camp, and C.C. Russ, "Advanced SCR ESD Protection Circuits for CMOS/SOI Nanotechnologies," in *Proceedings of the IEEE Custom Integrated Circuits Conference (CICC)*, pp. 481–488, 2005.
19. J. Li, J. Di Sarro, and R. Gauthier, "Design and Optimization of SCR Devices for On-chip ESD Protection in Advanced SOI CMOS Technologies," in *Proceedings of the IEEE International EOS/ESD Symposium*, pp. 1–8, 2012.
20. J. Li, J. Di Sarro, Y. Li, and R. Gauthier, "Investigation of SOI SCR Triggering and Current Sustaining under DC and TLP conditions," in *Proceedings of the IEEE International EOS/ESD Symposium*, pp. 1–6, 2013.

Index

Note: Locators followed by '*f*' and '*t*' refer to figures and tables, respectively.